ACT Summer Math Workbook

Essential Summer Learning Math Skills plus Two Complete ACT Math Practice Tests

By

Michael Smith & Reza Nazari

ACT Summer Math Workbook

Published in the United State of America By

The Math Notion

Web: WWW.MathNotion.Com

Email: info@Mathnotion.com

Copyright © 2020 by the Math Notion. All rights reserved. No part of this publication may be reproduced, stored in a retrieval system, or transmitted in any form or by any means, electronic, mechanical, photocopying, recording, scanning, or otherwise, except as permitted under Section 107 or 108 of the 1976 United States Copyright Ac, without permission of the author.

All inquiries should be addressed to the Math Notion.

About the Author

Michael Smith has been a math instructor for over a decade now. He holds a master's degree in Management. Since 2006, Michael has devoted his time to both teaching and developing exceptional math learning materials. As a Math instructor and test prep expert, Michael has worked with thousands of students. He has used the feedback of his students to develop a unique study program that can be used by students to drastically improve their math score fast and effectively.

- **SAT Math Practice Book**
- **PSAT Math Practice Book**
- **GRE Math Practice Book**
- **Accuplacer Math Practice Book**
- **Common Core Math Practice Book**
- **many Math Education Workbooks, Exercise Books and Study Guides**

As an experienced Math teacher, Mr. Smith employs a variety of formats to help students achieve their goals: He tutors online and in person, he teaches students in large groups, and he provides training materials and textbooks through his website and through Amazon.

You can contact Michael via email at:
info@Mathnotion.com

Prepare for the ACT Math test with a perfect workbook!

ACT Summer Math Workbook is a learning math workbook to prevent Summer learning loss. It helps students retain and strengthen their Math skills and provides a strong foundation for success. This workbook provides students with solid foundation to get a head starts on their upcoming school year.

ACT Summer Math Workbook is designed by top test prep experts to help students prepare for the ACT Math test. It provides test-takers with an in-depth focus on the math section of the test, helping them master the essential math skills that test-takers find the most troublesome. This is a prestigious resource for those who need an extra practice to succeed on the ACT Math test in the summer.

ACT Summer Math Workbook contains many exciting and unique features to help your student scores higher on the ACT Math test, including:

- Over 2,500 of standards-aligned math practice questions with answers
- Complete coverage of all Math concepts which students will need to ace the ACT test
- Content 100% aligned with the latest ACT test
- Written by ACT Math experts
- 2 full-length ACT Math practice tests (featuring new question types) with detailed answers

This Comprehensive Summer Workbook for the ACT Math is a perfect resource for those ACT Math test takers who want to review core content areas, brush-up in math, discover their strengths and weaknesses, and achieve their best scores on the ACT test.

WWW.MathNotion.COM

... So Much More Online!

- ✓ FREE Math Lessons

- ✓ More Math Learning Books!

- ✓ Mathematics Worksheets

- ✓ Online Math Tutors

For a PDF Version of This Book

Please Visit WWW.MathNotion.com

contents

Chapter 1: Integers and Number Theory **13**

Rounding 14

Whole Number Addition and Subtraction 15

Whole Number Multiplication and Division 16

Rounding and Estimates 17

Adding and Subtracting Integers 18

Multiplying and Dividing Integers 19

Order of Operations 20

Ordering Integers and Numbers 21

Integers and Absolute Value 22

Factoring Numbers 23

Greatest Common Factor 24

Least Common Multiple 25

Answers of Worksheets – Chapter 1 26

Chapter 2: Fractions and Decimals **29**

Simplifying Fractions 30

Adding and Subtracting Fractions 31

Multiplying and Dividing Fractions 32

Adding and Subtracting Mixed Numbers 33

Multiplying and Dividing Mixed Numbers 34

Adding and Subtracting Decimals 35

Multiplying and Dividing Decimals 36

Comparing Decimals 37

Rounding Decimals 38

Answers of Worksheets – Chapter 2 39

ACT Math Workbook

Chapter 3: Proportions, Ratios, and Percent 42

Simplifying Ratios 43
Proportional Ratios 44
Similarity and Ratios 45
Ratio and Rates Word Problems 46
Percentage Calculations 47
Percent Problems 48
Discount, Tax and Tip 49
Percent of Change 50
Simple Interest 51
Answers of Worksheets – Chapter 3 52

Chapter 4: Exponents and Radicals Expressions 55

Multiplication Property of Exponents 56
Zero and Negative Exponents 57
Division Property of Exponents 58
Powers of Products and Quotients 59
Negative Exponents and Negative Bases 60
Scientific Notation 61
Square Roots 62
Simplifying Radical Expressions 63
Multiplying Radical Expressions 64
Simplifying Radical Expressions Involving Fractions 65
Adding and Subtracting Radical Expressions 66
Answers of Worksheets – Chapter 4 67

Chapter 5: Algebraic Expressions 72

Simplifying Variable Expressions 73
Simplifying Polynomial Expressions 74
Translate Phrases into an Algebraic Statement 75
The Distributive Property 76
Evaluating One Variable Expressions 77
Evaluating Two Variables Expressions 78

Combining like Terms 79
Answers of Worksheets – Chapter 5 80

Chapter 6: Equations and Inequalities 82

One–Step Equations 83
Multi–Step Equations 84
Graphing Single–Variable Inequalities 85
One–Step Inequalities 86
Multi-Step Inequalities 87
Systems of Equations 88
Systems of Equations Word Problems 89
Answers of Worksheets – Chapter 6 90

Chapter 7: Linear Functions 93

Finding Slope 94
Graphing Lines Using Line Equation 95
Writing Linear Equations 96
Graphing Linear Inequalities 97
Finding Midpoint 98
Finding Distance of Two Points 99
Answers of Worksheets – Chapter 7 100

Chapter 8: Polynomials 103

Writing Polynomials in Standard Form 104
Simplifying Polynomials 105
Adding and Subtracting Polynomials 106
Multiplying Monomials 107
Multiplying and Dividing Monomials 108
Multiplying a Polynomial and a Monomial 109
Multiplying Binomials 110
Factoring Trinomials 111
Operations with Polynomials 112
Answers of Worksheets – Chapter 8 113

Chapter 9: Complex Numbers 117

Adding and Subtracting Complex Numbers 118
Multiplying and Dividing Complex Numbers 119
Graphing Complex Numbers..120
Rationalizing Imaginary Denominators 121
Answers of Worksheets – Chapter 9..122

Chapter 10: Functions Operations and Quadratic 123

Evaluating Function 124
Adding and Subtracting Functions 125
Multiplying and Dividing Functions 126
Composition of Functions 127
Quadratic Equation 128
Solving Quadratic Equations 129
Quadratic Formula and the Discriminant 130
Quadratic Inequalities 131
Graphing Quadratic Functions 132
Domain and Range of Radical Functions 133
Solving Radical Equations 134
Answers of Worksheets – Chapter 10..135

Chapter 11: Sequences and Series.. 140

Arithmetic Sequences 141
Geometric Sequences...142
Comparing Arithmetic and Geometric Sequences 143
Finite Geometric Series..144
Infinite Geometric Series...145
Answers of Worksheets – Chapter 11..146

Chapter 12: Logarithms 149

Rewriting Logarithms..150
Evaluating Logarithms..151
Properties of Logarithms 152
Natural Logarithms...153
Exponential Equations and Logarithms 154

Solving Logarithmic Equations 155

Answers of Worksheets – Chapter 12 156

Chapter 13: Geometry and Solid Figures 159

Angles... 160

Pythagorean Relationship 161

Triangles... 162

Polygons 163

Trapezoids 164

Circles 165

Cubes 166

Rectangular Prism 167

Cylinder.. 168

Pyramids and Cone.. 169

Answers of Worksheets – Chapter 13 170

Chapter 14: Trigonometric Functions 173

Trig Ratios of General Angles.. 174

Sketch Each Angle in Standard Position 175

Finding Co-terminal Angles and Reference Angles 176

Angles and Angle Measure 177

Evaluating Trigonometric Functions 178

Missing Sides and Angles of a Right Triangle 179

Arc Length and Sector Area 180

Answers of Worksheets – Chapter 14 181

Chapter 15: Statistics and Probability 183

Mean and Median.. 184

Mode and Range 185

Times Series 186

Stem–and–Leaf Plot.. 187

Pie Graph 188

Probability Problems... 189

Factorials 190

Combinations and Permutations 191
Answers of Worksheets – Chapter 15 192

ACT Math Test Review 195

ACT Math Practice Test Answer Sheets 197
ACT Practice Test 1 199
ACT Practice Test 2 217

Answers and Explanations 233

Answer Key 233
Practice Tests 1 235
Practice Tests 2 245

Chapter 1:
Integers and Number Theory

Topics that you will practice in this chapter:

- ✓ Rounding
- ✓ Whole Number Addition and Subtraction
- ✓ Whole Number Multiplication and Division
- ✓ Rounding and Estimates
- ✓ Adding and Subtracting Integers
- ✓ Multiplying and Dividing Integers
- ✓ Order of Operations
- ✓ Ordering Integers and Numbers
- ✓ Integers and Absolute Value
- ✓ Factoring Numbers
- ✓ Greatest Common Factor (GCF)
- ✓ Least Common Multiple (LCM)

"Wherever there is number, there is beauty." –Proclus

Rounding

✎ Round each number to the nearest ten.

1) 52 = ____   5) 42 = ____   9) 48 = ____

2) 89 = ____   6) 78 = ____   10) 21 = ____

3) 34 = ____   7) 121 = ____   11) 134 = ____

4) 66 = ____   8) 91 = ____   12) 157 = ____

✎ Round each number to the nearest hundred.

13) 148 = ____   17) 522 = ____   21) 780 = ____

14) 368 = ____   18) 169 = ____   22) 833 = ____

15) 619 = ____   19) 491 = ____   23) 498 = ____

16) 194 = ____   20) 717 = ____   24) 947 = ____

✎ Round each number to the nearest thousand.

25) 3,325 = ____   29) 6,075 = ____   33) 75,952 = ____

26) 2,598 = ____   30) 36,893 = ____   34) 95,250 = ____

27) 4,099 = ____   31) 52,199 = ____   35) 78,680 = ____

28) 5,808 = ____   32) 80,958 = ____   36) 97,869 = ____

Whole Number Addition and Subtraction

✎ Find the sum or subtract.

1) 1,982 + 895 = _____

2) 3,658 − 1,254 = _____

3) 582.54 − 321.45 = _____

4) 1,254 + 852.98 = _____

5) 1,125 + 859.35 = _____

6) 857.26 + 989.15 = _____

7) 254.35 + 123.89 = _____

8) 3,257.5 + 1,245.2 = _____

9) 7,322 − 895.9 = _____

10) 8,921.45 − 5,214.25 = _____

11) 2,321.25 + 1,984.99 = _____

12) 9,914.09 − 6,621.12 = _____

✎ Find the missing number.

13) $362.5 + \_\_\_ = 985.3$

14) $3,856 - \_\_\_ = 2,009.5$

15) $\_\_\_ - 985.1 = 1,450.9$

16) $2,785 - 1,234.12 = \_\_\_$

17) $999.9 + \_\_\_ = 1,234.6$

18) $5,758.8 - 3,758.85 = \_\_\_$

Whole Number Multiplication and Division

✎ **Calculate each product.**

1) 35
   × 43
   _____

2) 53.2
   × 12.5
   _____

3) 37.2
   × 16
   _____

4) 27.5
   × 26
   _____

5) 158.8
   × 15.4
   _____

6) 143.2
   × 15.5
   _____

✎ **Find the missing quotient.**

7) 600 ÷ 1.5 = _____

8) 780 ÷ 39 = _____

9) 390 ÷ 1.3 = _____

10) 900 ÷ 0.9 = _____

11) 156 ÷ 40 = _____

12) 112 ÷ 1.6 = _____

13) 660 ÷ 2.2 = _____

14) 400 ÷ 0.8 = _____

15) 2,040 ÷ 25.5 = _____

16) 9,360 ÷ 31.2 = _____

✎ **Calculate each problem.**

17) 560 ÷ 7 = N, N = __

18) 315 ÷ 4.5 = N, N = __

19) N ÷ 9 = 65, N = __

20) 24.6 × N = 147.6, N = __

21) 985 ÷ N = 1,970, N = __

22) N × 3.5 = 147, N = __

ACT Math Workbook

Rounding and Estimates

✏ **Estimate the sum by rounding each number to the nearest ten.**

1) 19 + 23 = _____

2) 72 + 31 = _____

3) 48 + 63 = _____

4) 44 + 86 = _____

5) 169 + 212 = _____

6) 650 + 323 = _____

7) 598 + 575 = _____

8) 1,586 + 3,355 = _____

✏ **Estimate the product by rounding each number to the nearest ten.**

9) 37 × 43 = _____

10) 12 × 31 = _____

11) 48 × 54 = _____

12) 17 × 33 = _____

13) 68 × 27 = _____

14) 91 × 21 = _____

15) 86 × 37 = _____

16) 96 × 42 = _____

✏ **Estimate the sum or product by rounding each number to the nearest ten.**

17) 28 × 16 = _____

18) 72 × 22 = _____

19) 85 + 64 = _____

20) 43 + 91 = _____

21) 64 × 39 = _____

22) 99 + 54 = _____

Adding and Subtracting Integers

✏ **Find each sum.**

1) $15 + (-35) =$

2) $(-28) + (-29) =$

3) $19 + (-27) =$

4) $57 + (-64) =$

5) $(-14) + (-19) + 64 =$

6) $54 + (-36) + 19 =$

7) $46 + (-30) + (-33) + 29 =$

8) $(-40) + (-70) + 28 + 55 =$

9) $60 + (-65) + (83 - 72) =$

10) $49 + (-55) + (90 - 67) =$

✏ **Find each difference.**

11) $(-32) - (-7) =$

12) $40 - (-12) =$

13) $(-60) - 56 =$

14) $27 - (-17) =$

15) $58 - (76 - 29) =$

16) $19 - (-14) - (-22) =$

17) $(39 + 15) - (-46) =$

18) $49 - 17 - (-13) =$

19) $85 - 45 - (-18) =$

20) $78 - (-35) - (-63) =$

21) $89 - (-11) - (-26) =$

22) $(19 - 50) - (-95) =$

23) $46 - 49 - (-87) =$

24) $120 - (98 + 24) - (-38) =$

25) $112 - (-102) + (-81) =$

26) $108 - (-42) + (-89) =$

Multiplying and Dividing Integers

🍃 Find each product.

1) $(-7) \times (-9) =$

2) $(-5) \times 6 =$

3) $10 \times (-15) =$

4) $(-9) \times (-25) =$

5) $(-7) \times (-12) \times 13 =$

6) $(15 - 4) \times (-11) =$

7) $25 \times (-4) \times (-5) =$

8) $(85 + 10) \times (-11) =$

9) $12 \times (-19 + 12) \times 5 =$

10) $(-15) \times (-18) \times (-20) =$

🍃 Find each quotient.

11) $85 \div (-5) =$

12) $(-90) \div (-15) =$

13) $(-121) \div (-11) =$

14) $99 \div (-33) =$

15) $(-114) \div 2 =$

16) $(-208) \div (-16) =$

17) $198 \div (-11) =$

18) $(-364) \div (-14) =$

19) $255 \div (-15) =$

20) $(-378) \div (18) =$

21) $(-184) \div (-8) =$

22) $-437 \div (-23) =$

23) $(-570) \div (-19) =$

24) $480 \div (-32) =$

25) $(-546) \div (-21) =$

26) $(486) \div (-54) =$

Order of Operations

Evaluate each expression.

1) $7 + (5 \times 8) =$

2) $16 - (6 \times 9) =$

3) $(17 \times 5) + 12 =$

4) $(24 - 12) - (11 \times 4) =$

5) $35 + (18 \div 3) =$

6) $(27 \times 3) \div 3 =$

7) $(88 \div 4) \times (-5) =$

8) $(9 \times 9) + (86 - 52) =$

9) $78 + (5 \times 12) + 14 =$

10) $(60 \times 4) \div (4 + 2) =$

11) $(-15) + (14 \times 4) + 18 =$

12) $(14 \times 5) - (56 \div 7) =$

13) $(7 \times 9 \div 3) - (32 + 21) =$

14) $(45 + 11 - 14) \times 2 - 15 =$

15) $(40 - 18 + 20) \times (75 \div 3) =$

16) $75 + (54 - (45 \div 9)) =$

17) $(12 + 15 - 24) + (44 \div 4) =$

18) $(78 - 19) + (27 - 10 + 7) =$

19) $(18 \times 3) + (17 \times 9) - 52 =$

20) $65 + 17 - (45 \times 2) + 40 =$

Ordering Integers and Numbers

✎ **Order each set of integers from least to greatest.**

1) $17, -15, -8, 0, 9$ ___, ___, ___, ___, ___, ___

2) $-14, -26, 17, 42, 39$ ___, ___, ___, ___, ___, ___

3) $32, -15, -69, 41, -80$ ___, ___, ___, ___, ___, ___

4) $-49, -65, 35, -21, 68$ ___, ___, ___, ___, ___, ___

5) $69, -32, 10, -45, 24$ ___, ___, ___, ___, ___, ___

6) $108, 76, -59, 87, -78$ ___, ___, ___, ___, ___, ___

✎ **Order each set of integers from greatest to least.**

7) $62, 98, -7, -19, -1$ ___, ___, ___, ___, ___, ___

8) $34, 35, -24, -46, 56$ ___, ___, ___, ___, ___, ___

9) $35, -96, -58, 17, -34$ ___, ___, ___, ___, ___, ___

10) $37, 12, -26, -13, 52$ ___, ___, ___, ___, ___, ___

11) $-12, 66, -18, -28, 54$ ___, ___, ___, ___, ___, ___

12) $-100, -85, -30, 5, 9$ ___, ___, ___, ___, ___, ___

Integers and Absolute Value

✎ **Write absolute value of each number.**

1) $|-19| =$

2) $|-32| =$

3) $|-50| =$

4) $|31| =$

5) $|57| =$

6) $|-76| =$

7) $|42| =$

8) $|101| =$

9) $|28| =$

10) $|-49| =$

11) $|-13|$

12) $|78| =$

13) $|100| =$

14) $|0| =$

15) $|-105| =$

16) $|-77| =$

17) $88 =$

18) $|-29| =$

19) $|112| =$

20) $|-120| =$

✎ **Evaluate the value.**

21) $|-5| - \dfrac{|-40|}{8} =$

22) $18 - |4 - 19| - |-15| =$

23) $\dfrac{|-72|}{9} \times |-9| =$

24) $\dfrac{|6 \times (-8)|}{3} \times \dfrac{|-21|}{7} =$

25) $|5 \times (-9)| + \dfrac{|-110|}{11} =$

26) $\dfrac{|-96|}{12} \times \dfrac{|-27|}{9} =$

27) $|-19 + 12| \times \dfrac{|-12 \times 13|}{7}$

28) $\dfrac{|-19 \times 6|}{3} \times |-11| =$

Factoring Numbers

✎ **List all positive factors of each number.**

1) 6

2) 21

3) 28

4) 26

5) 46

6) 45

7) 48

8) 50

9) 52

10) 63

11) 70

12) 72

13) 78

14) 80

15) 82

16) 88

17) 90

18) 93

19) 95

20) 96

21) 98

22) 102

23) 124

24) 125

Greatest Common Factor

Find the GCF for each number pair.

1) 6, 2

2) 8, 4

3) 5, 3

4) 6, 4

5) 7, 5

6) 8, 18

7) 14, 21

8) 6, 14

9) 9, 15

10) 4, 18

11) 14, 18

12) 25, 30

13) 27, 45

14) 36, 18

15) 9, 12

16) 11, 8

17) 28, 21

18) 56, 72

19) 34, 51

20) 6, 18, 27

21) 2, 9, 8

22) 10, 12, 24

23) 5, 14, 21

24) 72, 9, 18

Least Common Multiple

✏️ **Find the LCM for each number pair.**

1) 6, 5

2) 8, 18

3) 9, 15

4) 15, 20

5) 20, 25

6) 22, 33

7) 6, 28

8) 8, 14

9) 21, 28

10) 14, 28

11) 9, 30

12) 7, 12

13) 12, 36

14) 9, 54

15) 42, 21

16) 40, 16

17) 12, 42

18) 13, 11

19) 32, 72

20) 15, 27

21) 24, 44

22) 8, 12, 42

23) 2, 6, 11

24) 15, 25, 30

Answers of Worksheets – Chapter 1

Rounding

1) 50
2) 90
3) 30
4) 70
5) 40
6) 80
7) 120
8) 90
9) 50
10) 20
11) 130
12) 160
13) 100
14) 400
15) 600
16) 200
17) 500
18) 200
19) 500
20) 700
21) 800
22) 800
23) 500
24) 900
25) 3,000
26) 3,000
27) 4,000
28) 6,000
29) 6,000
30) 37,000
31) 52,000
32) 81,000
33) 76,000
34) 95,000
35) 79,000
36) 98,000

Whole Number Addition and Subtraction

1) 2,877
2) 2,404
3) 261.09
4) 2,106.98
5) 1,984.35
6) 1,846.41
7) 378.24
8) 4,502.7
9) 6,426.1
10) 3,707.2
11) 4,306.24
12) 3,292.97
13) 622.8
14) 1,846.5
15) 2,436
16) 1,550.88
17) 234.7
18) 1,999.95

Whole Number Multiplication and Division

1) 1,505
2) 665
3) 595.2
4) 715
5) 2,445.52
6) 2,219.6
7) 400
8) 20
9) 300
10) 1,000
11) 3.9
12) 70
13) 300
14) 500
15) 80
16) 300
17) 80
18) 70
19) 585
20) 6
21) 2
22) 42

Rounding and Estimates

1) 40
2) 100
3) 110
4) 130
5) 380
6) 970
7) 1,180
8) 4,950
9) 1,600
10) 300
11) 2,500
12) 600
13) 2,100
14) 1,800
15) 3,600
16) 4,200
17) 600
18) 1,400
19) 150
20) 130
21) 2,400
22) 150

ACT Math Workbook

Adding and Subtracting Integers

1) −20
2) −57
3) −8
4) −7
5) 31
6) 37
7) 12
8) −27
9) 6
10) 17
11) −25
12) 52
13) −116
14) 44
15) 11
16) 55
17) 100
18) 45
19) 58
20) 176
21) 126
22) 64
23) 84
24) 36
25) 133
26) 61

Multiplying and Dividing Integers

1) 63
2) −30
3) −150
4) 225
5) 1,092
6) −121
7) 500
8) −1,045
9) −420
10) −5,400
11) −17
12) 6
13) 11
14) −3
15) −57
16) 13
17) −18
18) 26
19) −17
20) −21
21) 23
22) 19
23) 30
24) −15
25) 26
26) −9

Order of Operations

1) 47
2) −38
3) 97
4) −32
5) 41
6) 27
7) −110
8) 115
9) 152
10) 40
11) 59
12) 62
13) −32
14) 69
15) 1,050
16) 124
17) 14
18) 83
19) 155
20) 32

Ordering Integers and Numbers

1) −15, −8, 0, 9, 17
2) −26, −14, 17, 39, 42
3) −80, −69, −15, 32, 41
4) −65, −49, −21, 35, 68
5) −45, −32, 10, 24, 69
6) −78, −59, 76, 87, 108
7) 98, 62, −1, −7, −19
8) 56, 35, 34, −24, −46
9) 35, 17, −34, −58, −96
10) 52, 37, 12, −13, −26
11) 66, 54, −12, −18, −28
12) 9, 5, −30, −85, −100

Integers and Absolute Value

1) 19
2) 32
3) 50
4) 31

ACT Math Workbook

5) 57	11) 13	17) 88	23) 72
6) 76	12) 78	18) 29	24) 48
7) 42	13) 100	19) 112	25) 55
8) 101	14) 0	20) 120	26) 24
9) 28	15) 105	21) 0	27) 156
10) 49	16) 77	22) -12	28) 418

Factoring Numbers

1) 1, 2, 3, 6
2) 1, 3, 7, 21
3) 1, 2, 4, 7, 14, 28
4) 1, 2, 13, 26
5) 1, 2, 23, 46
6) 1, 3, 5, 9, 15, 45
7) 1, 2, 3, 4, 6, 8, 12, 16, 24, 48
8) 1, 2, 5, 10, 25, 50
9) 1, 2, 4, 5, 13, 26, 52
10) 1, 3, 7, 9, 21, 63
11) 1, 2, 5, 7, 10, 14, 35, 70
12) 1, 2, 3, 4, 6, 8, 9, 12, 18 24, 36, 72
13) 1, 2, 3, 6, 13, 26, 39, 78
14) 1, 2, 4, 5, 8, 10, 16, 20, 40, 80
15) 1, 2, 41, 82
16) 1, 2, 4, 8, 11, 22, 44, 88
17) 1, 2, 3, 5, 6, 9, 10, 15, 18, 30, 45, 90
18) 1, 3, 31, 93
19) 1, 5, 19, 95
20) 1, 2, 3, 4, 6, 8, 12, 16, 24, 32, 48, 96
21) 1, 2, 7, 14, 49, 98
22) 1, 2, 3, 6, 17, 34, 51, 102
23) 1, 2, 4, 31, 62, 124
24) 1, 5, 25, 125

Greatest Common Factor

1) 2	7) 7	13) 9	19) 17
2) 4	8) 2	14) 18	20) 3
3) 1	9) 3	15) 3	21) 1
4) 2	10) 2	16) 1	22) 2
5) 1	11) 2	17) 7	23) 1
6) 2	12) 5	18) 8	24) 9

Least Common Multiple

1) 30	7) 84	13) 36	19) 288
2) 72	8) 56	14) 54	20) 135
3) 45	9) 84	15) 42	21) 264
4) 60	10) 28	16) 80	22) 168
5) 100	11) 90	17) 84	23) 66
6) 66	12) 84	18) 143	24) 150

Chapter 2:

Fractions and Decimals

Topics that you will practice in this chapter:

- ✓ Simplifying Fractions
- ✓ Adding and Subtracting Fractions
- ✓ Multiplying and Dividing Fractions
- ✓ Adding and Subtract Mixed Numbers
- ✓ Multiplying and Dividing Mixed Numbers
- ✓ Adding and Subtracting Decimals
- ✓ Multiplying and Dividing Decimals
- ✓ Comparing Decimals
- ✓ Rounding Decimals

"A Man is like a fraction whose numerator is what he is and whose denominator is what he thinks of himself. The larger the denominator, the smaller the fraction." −Tolstoy

Simplifying Fractions

✎ **Simplify each fraction to its lowest terms.**

1) $\dfrac{8}{16} =$

2) $\dfrac{28}{35} =$

3) $\dfrac{27}{36} =$

4) $\dfrac{70}{140} =$

5) $\dfrac{13}{52} =$

6) $\dfrac{38}{57} =$

7) $\dfrac{64}{80} =$

8) $\dfrac{21}{84} =$

9) $\dfrac{85}{170} =$

10) $\dfrac{120}{168} =$

11) $\dfrac{31}{124} =$

12) $\dfrac{48}{96} =$

13) $\dfrac{98}{112} =$

14) $\dfrac{99}{110} =$

15) $\dfrac{51}{153} =$

16) $\dfrac{40}{112} =$

17) $\dfrac{90}{225} =$

18) $\dfrac{44}{297} =$

19) $\dfrac{54}{279} =$

20) $\dfrac{320}{720} =$

21) $\dfrac{70}{560} =$

✎ **Find the answer for each problem.**

22) Which of the following fractions equal to $\dfrac{3}{7}$? _____

    A. $\dfrac{24}{63}$ B. $\dfrac{51}{109}$ C. $\dfrac{51}{119}$ D. $\dfrac{240}{630}$

23) Which of the following fractions equal to $\dfrac{7}{8}$? _____

    A. $\dfrac{182}{208}$ B. $\dfrac{175}{208}$ C. $\dfrac{182}{216}$ D. $\dfrac{49}{64}$

24) Which of the following fractions equal to $\dfrac{2}{9}$? _____

    A. $\dfrac{64}{126}$ B. $\dfrac{46}{207}$ C. $\dfrac{48}{207}$ D. $\dfrac{56}{208}$

Adding and Subtracting Fractions

✎ **Find the sum.**

1) $\dfrac{5x}{8} + \dfrac{3x}{8} =$

2) $\dfrac{x}{2} + \dfrac{x}{7} =$

3) $\dfrac{y}{3} + \dfrac{y}{4} =$

4) $\dfrac{3x}{8} + \dfrac{2x}{5} =$

5) $\dfrac{xy}{5} + \dfrac{2xy}{7} =$

6) $\dfrac{2x}{9} + \dfrac{4x}{9} =$

7) $\dfrac{a}{4} + \dfrac{2a}{3} =$

8) $\dfrac{2}{x} + \dfrac{4}{x} =$

9) $\dfrac{1}{a} + \dfrac{2}{b} =$

10) $\dfrac{3b}{5} + \dfrac{2b}{7} =$

11) $\dfrac{a}{y} + \dfrac{3a}{y} =$

12) $\dfrac{3}{x} + \dfrac{1}{2x} =$

✎ **Find the difference.**

13) $\dfrac{x}{3} - \dfrac{x}{6} =$

14) $\dfrac{2x}{5} - \dfrac{3x}{8} =$

15) $\dfrac{x}{7} - \dfrac{y}{7} =$

16) $\dfrac{2x}{7} - \dfrac{x}{6} =$

17) $\dfrac{5a}{9} - \dfrac{2a}{5} =$

18) $\dfrac{2ab}{3} - \dfrac{ab}{6} =$

19) $\dfrac{1}{x} - \dfrac{1}{3x} =$

20) $\dfrac{4}{y} - \dfrac{3}{4y} =$

21) $\dfrac{5}{x} - \dfrac{2y}{xy} =$

22) $\dfrac{8}{ab} - \dfrac{5}{3ab} =$

23) $\dfrac{2a}{y} - \dfrac{a}{3y} =$

24) $\dfrac{5}{b} - \dfrac{2}{3b} =$

25) $\dfrac{2a}{b} - \dfrac{a}{b} =$

26) $\dfrac{3}{a} - \dfrac{2}{b} =$

27) $\dfrac{4a}{b} - \dfrac{2a}{3b} =$

28) $\dfrac{6}{xy} - \dfrac{7}{2xy} =$

29) $\dfrac{2}{a} - \dfrac{1}{4a} =$

30) $\dfrac{2a}{3b} - \dfrac{4a}{9b} =$

Multiplying and Dividing Fractions

✎ **Find the value of each expression in lowest terms.**

1) $\dfrac{3}{a} \times \dfrac{5}{3} =$

2) $\dfrac{2}{3b} \times \dfrac{9}{2} =$

3) $\dfrac{a}{15} \times \dfrac{5}{2a} =$

4) $\dfrac{x}{3a} \times \dfrac{9a}{6x} =$

5) $\dfrac{x}{12} \times \dfrac{6}{y} =$

6) $\dfrac{7}{x} \times \dfrac{x}{14} =$

7) $\dfrac{10}{3a} \times \dfrac{6}{20} =$

8) $\dfrac{4a}{b} \times \dfrac{2b}{5} =$

9) $\dfrac{2ab}{7} \times \dfrac{14}{6ab} =$

10) $\dfrac{4a}{5b} \times \dfrac{15}{2a} =$

11) $\dfrac{ab}{21} \times \dfrac{7}{a} =$

12) $\dfrac{a}{cd} \times \dfrac{2bc}{a} =$

✎ **Find the value of each expression in lowest terms.**

13) $\dfrac{a}{2} \div \dfrac{a}{4} =$

14) $\dfrac{b}{3} \div \dfrac{b}{9} =$

15) $\dfrac{a}{b} \div \dfrac{3}{b} =$

16) $\dfrac{2a}{15} \div \dfrac{4a}{5} =$

17) $\dfrac{1}{a} \div \dfrac{b}{3a} =$

18) $\dfrac{4a}{3b} \div \dfrac{a}{2b} =$

19) $\dfrac{a}{8} \div \dfrac{3a}{16} =$

20) $\dfrac{3b}{20} \div \dfrac{6b}{15a} =$

21) $\dfrac{x}{12y} \div \dfrac{2x}{9y} =$

22) $\dfrac{25}{x} \div \dfrac{50}{2x} =$

23) $\dfrac{16}{5ab} \div \dfrac{32}{ab} =$

24) $\dfrac{7a}{b} \div \dfrac{8a}{b} =$

25) $\dfrac{5}{x} \div \dfrac{3y}{x} =$

26) $\dfrac{2a}{21} \div \dfrac{a}{14} =$

27) $\dfrac{ab}{x} \div \dfrac{a}{x} =$

28) $\dfrac{6}{a} \div \dfrac{3b}{2a} =$

29) $\dfrac{9}{16a} \div \dfrac{3}{8ab} =$

30) $\dfrac{24}{xy} \div \dfrac{12}{y} =$

Adding and Subtracting Mixed Numbers

✎ **Find the sum.**

1) $3\frac{5}{6} + 2\frac{1}{3} =$

2) $4\frac{2}{5} + 1\frac{1}{5} =$

3) $5\frac{1}{8} + 6\frac{3}{4} =$

4) $2\frac{2}{3} + 3\frac{1}{2} =$

5) $3\frac{4}{5} + 3\frac{2}{15} =$

6) $8\frac{1}{16} + 3\frac{3}{8} =$

7) $4\frac{3}{5} + 4\frac{1}{6} =$

8) $7\frac{3}{4} + 3\frac{5}{6} =$

9) $8\frac{5}{6} + 2\frac{2}{7} =$

10) $11\frac{3}{16} + 3\frac{5}{24} =$

✎ **Find the difference.**

11) $3\frac{3}{4} - 2\frac{1}{4} =$

12) $5\frac{1}{7} - 3\frac{1}{7} =$

13) $4\frac{1}{3} - 1\frac{1}{9} =$

14) $7\frac{1}{6} - 3\frac{1}{12} =$

15) $6\frac{1}{3} - 2\frac{5}{18} =$

16) $8\frac{1}{4} - 5\frac{1}{8} =$

17) $9\frac{1}{2} - 6\frac{1}{5} =$

18) $11\frac{7}{15} - 8\frac{1}{30} =$

19) $12\frac{3}{5} - 7\frac{2}{7} =$

20) $18\frac{1}{8} - 14\frac{3}{16} =$

21) $12\frac{2}{3} - 11\frac{7}{15} =$

22) $3\frac{1}{5} - 1\frac{1}{2} =$

23) $14\frac{3}{5} - 6\frac{4}{5} =$

24) $17\frac{1}{4} - 14\frac{8}{9} =$

25) $24\frac{3}{9} - 15\frac{1}{18} =$

26) $28\frac{3}{7} - 19\frac{5}{6} =$

Multiplying and Dividing Mixed Numbers

✏️ **Find the product.**

1) $2\frac{1}{3} \times 4\frac{1}{2} =$

2) $4\frac{1}{5} \times 2\frac{1}{3} =$

3) $7\frac{2}{3} \times 3\frac{3}{5} =$

4) $9\frac{2}{7} \times 3\frac{1}{8} =$

5) $5\frac{4}{11} \times 4\frac{1}{3} =$

6) $7\frac{3}{8} \times 5\frac{4}{9} =$

7) $9\frac{2}{3} \times 11\frac{5}{6} =$

8) $8\frac{3}{5} \times 7\frac{4}{9} =$

9) $5\frac{1}{9} \times 9\frac{5}{8} =$

10) $10\frac{2}{7} \times 2\frac{5}{8} =$

✏️ **Find the quotient.**

11) $2\frac{1}{8} \div 1\frac{3}{8} =$

12) $4\frac{1}{6} \div 2\frac{1}{3} =$

13) $7\frac{1}{3} \div 3\frac{3}{4} =$

14) $4\frac{5}{8} \div 1\frac{1}{2} =$

15) $6\frac{5}{12} \div 4\frac{1}{6} =$

16) $5\frac{7}{18} \div 5\frac{1}{6} =$

17) $6\frac{5}{21} \div 2\frac{3}{7} =$

18) $8\frac{1}{7} \div 8\frac{1}{14} =$

19) $10\frac{1}{4} \div 3\frac{2}{5} =$

20) $15\frac{1}{3} \div 5\frac{2}{9} =$

21) $12\frac{1}{3} \div 6\frac{1}{2} =$

22) $18\frac{1}{9} \div 18\frac{1}{6} =$

23) $10\frac{3}{4} \div 5\frac{2}{5} =$

24) $11\frac{1}{3} \div 8\frac{4}{5} =$

25) $9\frac{1}{6} \div 3\frac{2}{7} =$

26) $7\frac{1}{3} \div 3\frac{7}{11} =$

Adding and Subtracting Decimals

✎ **Add and subtract decimals.**

1) 52.18 − 21.27

2) 49.34 + 25.24

3) 48.60 + 35.75

4) 65.84 − 35.49

5) 54.57 + 18.37

6) 90.45 − 28.75

7) 98.12 − 45.55

8) 48.99 + 57.67

9) 158.05 − 78.98

✎ **Find the missing number.**

10) ___ + 4.9 = 6.5

11) 5.15 + ___ = 6.43

12) 8.09 + ___ = 11.84

13) 8.88 − ___ = 6.78

14) ___ − 1.59 = 3.71

15) ___ − 19.98 = 8.17

16) 38.89 + ___ = 41.32

17) ___ − 35.99 = 1.80

18) ___ + 39.08 = 41.36

19) 98.98 + ___ = 123.68

Multiplying and Dividing Decimals

Find the product.

1) $0.6 \times 0.8 =$

2) $2.5 \times 0.9 =$

3) $0.87 \times 0.4 =$

4) $0.15 \times 0.75 =$

5) $0.95 \times 0.7 =$

6) $1.57 \times 0.9 =$

7) $5.85 \times 1.3 =$

8) $12.5 \times 4.5 =$

9) $19.8 \times 7.32 =$

10) $85.1 \times 1.5 =$

11) $79.5 \times 11.2 =$

12) $86.9 \times 21.5 =$

Find the quotient.

13) $3.25 \div 10 =$

14) $24.5 \div 100 =$

15) $3.9 \div 3 =$

16) $91.2 \div 0.6 =$

17) $29.2 \div 0.4 =$

18) $38.7 \div 9 =$

19) $297.8 \div 1,000 =$

20) $53.55 \div 0.7 =$

21) $345.45 \div 0.1 =$

22) $70.27 \div 0.25 =$

23) $28.968 \div 0.3 =$

24) $86.34 \div 0.06 =$

Comparing Decimals

✎ Write the correct comparison symbol (>, < or =).

1) 0.80 ☐ 0.080

2) 0.086 ☐ 0.86

3) 7.090 ☐ 7.09

4) 3.25 ☐ 3.06

5) 4.09 ☐ 0.490

6) 6.06 ☐ 6.6

7) 6.08 ☐ 6.080

8) 4.05 ☐ 4.2

9) 12.35 ☐ 12.198

10) 0.957 ☐ 0.0957

11) 25.24 ☐ 25.240

12) 0.742 ☐ 0.752

13) 14.09 ☐ 14.10

14) 17.45 ☐ 17.154

15) 11.44 ☐ 11.439

16) 15.41 ☐ 15.410

17) 21.43 ☐ 21.043

18) 8.098 ☐ 8.90

19) 16.044 ☐ 16.040

20) 32.35 ☐ 32.350

Rounding Decimals

✎ **Round each decimal to the nearest whole number.**

1) 56.27   3) 18.32   5) 7.90

2) 5.9   4) 4.8   6) 57.7

✎ **Round each decimal to the nearest tenth.**

7) 42.785   9) 96.586   11) 27.198

8) 15.224   10) 101.78   12) 96.87

✎ **Round each decimal to the nearest hundredth.**

13) 9.648   15) 89.2882   17) 68.229

14) 27.819   16) 120.912   18) 85.642

✎ **Round each decimal to the nearest thousandth.**

19) 19.88486   21) 145.9322   23) 189.0991

20) 46.72611   22) 210.1581   24) 121.76798

Answers of Worksheets – Chapter 2

Simplifying Fractions

1) $\frac{1}{2}$
2) $\frac{4}{5}$
3) $\frac{3}{4}$
4) $\frac{1}{2}$
5) $\frac{1}{4}$
6) $\frac{2}{3}$
7) $\frac{4}{5}$
8) $\frac{1}{4}$
9) $\frac{1}{2}$
10) $\frac{5}{7}$
11) $\frac{1}{4}$
12) $\frac{1}{2}$
13) $\frac{7}{8}$
14) $\frac{9}{10}$
15) $\frac{1}{3}$
16) $\frac{5}{14}$
17) $\frac{2}{5}$
18) $\frac{4}{27}$
19) $\frac{6}{31}$
20) $\frac{4}{9}$
21) $\frac{1}{8}$
22) C
23) A
24) B

Adding and Subtracting Fractions

1) $\frac{8x}{8} = x$
2) $\frac{9x}{14}$
3) $\frac{7x}{12}$
4) $\frac{31x}{40}$
5) $\frac{17xy}{35}$
6) $\frac{2x}{3}$
7) $\frac{11a}{12}$
8) $\frac{6}{x}$
9) $\frac{a+2b}{ab}$
10) $\frac{31b}{35}$
11) $\frac{4a}{y}$
12) $\frac{7}{2x}$
13) $\frac{x}{6}$
14) $\frac{x}{40}$
15) $\frac{x-y}{7}$
16) $\frac{5x}{42}$
17) $\frac{7a}{45}$
18) $\frac{ab}{2}$
19) $\frac{2}{3x}$
20) $\frac{13}{4y}$
21) $\frac{3}{x}$
22) $\frac{19}{3ab}$
23) $\frac{5a}{3y}$
24) $\frac{13}{3b}$
25) $\frac{a}{b}$
26) $\frac{3b-2a}{ab}$
27) $\frac{10a}{3b}$
28) $\frac{5}{2xy}$
29) $\frac{7}{4a}$
30) $\frac{2a}{9b}$

Multiplying and Dividing Fractions

1) $\frac{5}{a}$
2) $\frac{3}{b}$
3) $\frac{1}{6}$
4) $\frac{1}{2}$
5) $\frac{x}{2y}$
6) $\frac{1}{2}$
7) $\frac{1}{a}$
8) $\frac{8a}{5}$
9) $\frac{2}{3}$

ACT Math Workbook

10) $\frac{6}{b}$

11) $\frac{b}{3}$

12) $\frac{2b}{d}$

13) 2

14) 3

15) $\frac{a}{3}$

16) $\frac{1}{6}$

17) $\frac{3}{b}$

18) $\frac{8}{3}$

19) $\frac{2}{3}$

20) $\frac{3a}{8}$

21) $\frac{3}{8}$

22) 1

23) $\frac{1}{10}$

24) $\frac{7}{8}$

25) $\frac{5}{3y}$

26) $\frac{4}{3}$

27) b

28) $\frac{4}{b}$

29) $\frac{3b}{2}$

30) $\frac{2}{x}$

Adding and Subtracting Mixed Numbers

1) $6\frac{1}{6}$

2) $5\frac{3}{5}$

3) $11\frac{7}{8}$

4) $6\frac{1}{6}$

5) $6\frac{14}{15}$

6) $11\frac{7}{16}$

7) $8\frac{23}{30}$

8) $11\frac{7}{12}$

9) $11\frac{5}{42}$

10) $14\frac{19}{48}$

11) $1\frac{1}{2}$

12) 2

13) $3\frac{2}{9}$

14) $4\frac{1}{12}$

15) $4\frac{1}{18}$

16) $3\frac{1}{8}$

17) $3\frac{3}{10}$

18) $3\frac{13}{30}$

19) $5\frac{11}{35}$

20) $3\frac{15}{16}$

21) $1\frac{1}{5}$

22) $1\frac{7}{10}$

23) $7\frac{4}{5}$

24) $2\frac{13}{36}$

25) $9\frac{5}{18}$

26) $8\frac{25}{42}$

Multiplying and Dividing Mixed Numbers

1) $10\frac{1}{2}$

2) $9\frac{4}{5}$

3) $27\frac{3}{5}$

4) $29\frac{1}{56}$

5) $23\frac{8}{33}$

6) $40\frac{11}{72}$

7) $144\frac{7}{18}$

8) $64\frac{1}{45}$

9) $49\frac{7}{36}$

10) 27

11) $1\frac{6}{11}$

12) $1\frac{11}{14}$

13) $1\frac{43}{45}$

14) $3\frac{1}{12}$

15) $1\frac{27}{50}$

16) $1\frac{4}{93}$

17) $2\frac{29}{51}$

18) $1\frac{1}{113}$

19) $3\frac{1}{68}$

20) $2\frac{44}{47}$

21) $1\frac{35}{39}$

ACT Math Workbook

22) $\frac{326}{327}$

23) $1\frac{107}{108}$

24) $1\frac{19}{66}$

25) $2\frac{109}{138}$

26) $2\frac{1}{60}$

Adding and Subtracting Decimals

1) 30.91
2) 74.58
3) 84.35
4) 30.35
5) 72.94
6) 61.7
7) 52.57
8) 106.66
9) 79.07
10) 1.6
11) 1.28
12) 3.75
13) 2.1
14) 5.3
15) 28.15
16) 2.43
17) 37.79
18) 2.28
19) 24.7

Multiplying and Dividing Decimals

1) 0.48
2) 2.25
3) 0.348
4) 0.1125
5) 0.665
6) 1.413
7) 7.605
8) 56.25
9) 144.936
10) 127.65
11) 890.4
12) 1,868.35
13) 0.325
14) 0.245
15) 1.3
16) 152
17) 73
18) 4.3
19) 0.2978
20) 76.5
21) 3,454.5
22) 281.08
23) 96.56
24) 1,439

Comparing Decimals

1) >
2) <
3) =
4) >
5) >
6) <
7) =
8) <
9) >
10) >
11) =
12) <
13) <
14) >
15) >
16) =
17) >
18) <
19) >
20) =

Rounding Decimals

1) 56
2) 6
3) 18
4) 5
5) 8
6) 58
7) 42.8
8) 15.2
9) 96.6
10) 101.8
11) 27.2
12) 96.9
13) 9.65
14) 27.82
15) 89.29
16) 120.91
17) 68.23
18) 85.64
19) 19.885
20) 46.726
21) 145.932
22) 210.158
23) 189.099
24) 121.768

Chapter 3: Proportions, Ratios, and Percent

Topics that you will practice in this chapter:

- ✓ Simplifying Ratios
- ✓ Proportional Ratios
- ✓ Similarity and Ratios
- ✓ Ratio and Rates Word Problems
- ✓ Percentage Calculations
- ✓ Percent Problems
- ✓ Discount, Tax and Tip
- ✓ Percent of Change
- ✓ Simple Interest

Without mathematics, there's nothing you can do. Everything around you is mathematics. Everything around you is numbers." – Shakuntala Devi

Simplifying Ratios

🍂 **Reduce each ratio.**

1) 15 : 20 = ___ : ___

2) 9 : 90 = ___ : ___

3) 24 : 42 = ___ : ___

4) 7 : 21 = ___ : ___

5) 11 : 110 ___ : ___

6) 8 : 64 = ___ : ___

7) 18 : 72 = ___ : ___

8) 10 : 25 = ___ : ___

9) 7 : 42 = ___ : ___

10) 49 : 63 = ___ : ___

11) 12 : 18 ___ : ___

12) 35 : 10 ___ : ___

13) 150 : 15 ___ : ___

14) 2.4 : 3.2 ___ : ___

15) 7 : 56 = ___ : ___

16) 45 : 63 ___ : ___

17) 77 : 99 ___ : ___

18) 39 : 13 ___ : ___

19) 15 : 45 ___ : ___

20) 84 : 12 ___ : ___

21) 25 : 5 ___ : ___

22) 70 : 56 ___ : ___

23) 70 : 140 ___ : ___

24) 1.2 : 36 ___ : ___

🍂 **Write each ratio as a fraction in simplest form.**

25) 7 : 14 =

26) 27 : 45 =

27) 24 : 56 =

28) 16 : 48 =

29) 22 : 66 =

30) 21 : 98 =

31) 34 : 68 =

32) 6 : 30 =

33) 35 : 84 =

34) 12 : 54 =

35) 88 : 104 =

36) 36 : 81 =

37) 1.5 : 18 =

38) 4.5 : 16.5 =

39) 5 : 75 =

40) 3.1 : 12.4 =

41) 1.6 : 6.4 =

42) 0.25 : 1.25 =

43) 8.8 : 16.4 =

44) 0.75 : 6.75 =

45) 1.8 : 3 =

Proportional Ratios

✎ **Fill in the blanks; Calculate each proportion.**

1) $3 : 8 = \_\_ : 32$

2) $1 : 2 = 45 : \_\_$

3) $1 : 11 = \_\_ : 55$

4) $9 : 12 = 18 : \_\_$

5) $9 : 7 = 81 : \_\_$

6) $2 : 8 = \_\_ : 56$

7) $2.3 : 1.2 = \_\_ : 12$

8) $0.5 : 2 = \_\_ : 32$

9) $1.6 : 2 = \_\_ : 60$

10) $2.5 : 4.5 = \_\_ : 90$

11) $3.8 : 7.1 = 7.6 : \_\_$

12) $5.5 : 6 = 16.5 : \_\_$

✎ **State if each pair of ratios form a proportion.**

13) $\frac{5}{12}$ and $\frac{15}{36}$

14) $\frac{2}{4}$ and $\frac{18}{36}$

15) $\frac{7}{8}$ and $\frac{28}{32}$

16) $\frac{3}{8}$ and $\frac{27}{64}$

17) $\frac{1}{14}$ and $\frac{5}{65}$

18) $\frac{7}{11}$ and $\frac{70}{100}$

19) $\frac{12}{15}$ and $\frac{48}{60}$

20) $\frac{3}{17}$ and $\frac{36}{204}$

21) $\frac{1.2}{1.5}$ and $\frac{1.44}{22.5}$

22) $\frac{1.3}{1.1}$ and $\frac{3.9}{33}$

23) $\frac{0.7}{0.9}$ and $\frac{6.3}{8.1}$

24) $\frac{2.4}{3.2}$ and $\frac{48}{64}$

✎ **Calculate each proportion.**

25) $\frac{14}{16} = \frac{21}{x}, x = \_\_\_$

26) $\frac{3}{28} = \frac{42}{x}, x = \_\_\_$

27) $\frac{19}{5} = \frac{38}{x}, x = \_\_\_$

28) $\frac{3}{10} = \frac{x}{140}, x = \_\_\_$

29) $\frac{4}{9} = \frac{x}{108}, x = \_\_\_$

30) $\frac{7}{32} = \frac{21}{x}, x = \_\_\_$

31) $\frac{9}{8} = \frac{108}{x}, x = \_\_\_$

32) $\frac{12}{17} = \frac{48}{x}, x = \_\_\_$

33) $\frac{1.4}{5} = \frac{x}{30}, x = \_\_\_$

34) $\frac{1.6}{12} = \frac{x}{60}, x = \_\_\_$

35) $\frac{3.5}{15} = \frac{x}{315}, x = \_\_\_$

36) $\frac{4.7}{2.5} = \frac{x}{50}, x = \_\_\_$

Similarity and Ratios

✎ **Each pair of figures is similar. Find the missing side.**

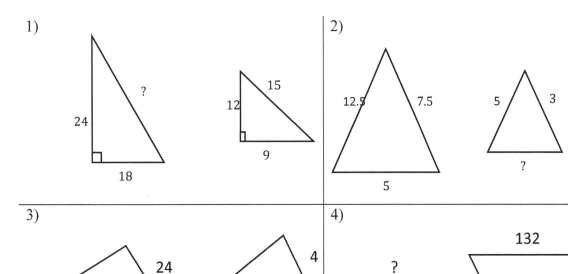

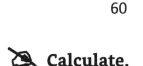

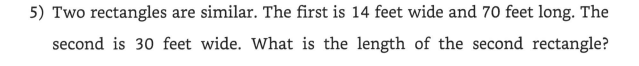

✎ **Calculate.**

5) Two rectangles are similar. The first is 14 feet wide and 70 feet long. The second is 30 feet wide. What is the length of the second rectangle? _____

6) Two rectangles are similar. One is 3.2 meters by 15 meters. The longer side of the second rectangle is 42 meters. What is the other side of the second rectangle? _____

7) A building casts a shadow 24 ft long. At the same time a girl 10 ft tall casts a shadow 6 ft long. How tall is the building? _____

8) The scale of a map of Texas is 8 inches: 52 miles. If you measure the distance from Dallas to Martin County as 28.8 inches, approximately how far is Martin County from Dallas? _____

Ratio and Rates Word Problems

✎ **Find the answer for each word problem.**

1) Mason has 32 red cards and 40 green cards. What is the ratio of Mason's red cards to his green cards? _____

2) In a party, 24 soft drinks are required for every 42 guests. If there are 378 guests, how many soft drinks is required? _____

3) In Mason's class, 54 of the students are tall and 30 are short. In Michael's class 126 students are tall and 70 students are short. Which class has a higher ratio of tall to short students? _____

4) The price of 4 apples at the Quick Market is $3.65. The price of 6 of the same apples at Walmart is $4.25. Which place is the better buy? _____

5) The bakers at a Bakery can make 90 bagels in 3 hours. How many bagels can they bake in 17 hours? What is that rate per hour? _____

6) You can buy 8 cans of green beans at a supermarket for $5.60. How much does it cost to buy 56 cans of green beans? _____

7) The ratio of boys to girls in a class is 4: 7. If there are 16 boys in the class, how many girls are in that class? _____

8) The ratio of red marbles to blue marbles in a bag is 3: 4. If there are 42 marbles in the bag, how many of the marbles are red? _____

Percentage Calculations

✏️ **Calculate the given percent of each value.**

1) 3% of 60 = ____

2) 20% of 80 = ____

3) 25% of 80 = ____

4) 24% of 50 = ____

5) 18% of 150 = ____

6) 70% of 35 = ____

7) 15% of 28 = ____

8) 32% of 300 = ____

9) 54% of 80 = ____

10) 10% of 610 = ____

11) 35% of 520 = ____

12) 64% of 110 = ____

13) 44% of 200 = ____

14) 28% of 94 = ____

15) 30% of 85 = ____

16) 68% of 102 = ____

17) 45% of 160 = ____

18) 55% of 220 = ____

✏️ **Calculate the percent of each given value.**

19) ____% of 18 = 9

20) ____% of 50 = 40

21) ____% of 140 = 7

22) ____% of 158 = 39.5

23) ____% of 75 = 9.375

24) ____% of 45 = 11.25

25) ____% of 90 = 22.5

26) ____% of 650 = 19.5

27) ____% of 480 = 24

28) ____% of 400 = 57.32

✏️ **Calculate each percent problem.**

29) A Cinema has 132 seats. 92 seats were sold for the current movie. What percent of seats are empty? ____ %

30) There are 52 boys and 68 girls in a class. 55.00% of the students in the class take the bus to school. How many students do not take the bus to school? ____

Percent Problems

Calculate each problem.

1) 30 is what percent of 60? ___%

2) 32 is what percent of 80? ___%

3) 72 is what percent of 45? ___%

4) 8 is what percent of 200? ___%

5) 9 is what percent of 600? ___%

6) 30 is what percent of 500? ___%

7) 70 is what percent of 350? ___%

8) 44 is what percent of 550? ___%

9) 270 is what percent of 900? ___%

10) 180 is what percent of 720? ___%

11) 37.5 is what percent of 75? ___%

12) 27.5 is what percent of 55? ___%

13) 60 is what percent of 750? ___%

14) 22.5 is what percent of 18? ___%

15) 36 is what percent of 24? ___%

16) 18 is what percent of 60? ___%

17) 140 is what percent of 280? ___%

18) 128 is what percent of 40? ___%

Calculate each percent word problem.

19) There are 32 employees in a company. On a certain day, 24 were present. What percent showed up for work? ___%

20) A metal bar weighs 36 ounces. 40% of the bar is gold. How many ounces of gold are in the bar? ___

21) A crew is made up of 12 women; the rest are men. If 20% of the crew are women, how many people are in the crew? ___

22) There are 32 students in a class and 8 of them are girls. What percent are boys? ___%

23) The Royals softball team played 310 games and won 248 of them. What percent of the games did they lose? ___%

Discount, Tax and Tip

✎ **Find the selling price of each item.**

1) Original price of a computer: $450
   Tax: 8% Selling price: $_____

2) Original price of a laptop: $240
   Tax: 4% Selling price: $_____

3) Original price of a sofa: $900
   Tax: 12%   Selling price: $_____

4) Original price of a car: $10,400
   Tax: 2.5%   Selling price: $_____

5) Original price of a Table: $400
   Tax: 3% Selling price: $_____

6) Original price of a house: $360,000
   Tax: 2.8%   Selling price: $_____

7) Original price of a tablet: $150
   Discount: 24%   Selling price: $____

8) Original price of a chair: $180
   Discount: 20%   Selling price: $____

9) Original price of a book: $80
   Discount: 30%   Selling price: $____

10) Original price of a cellphone: $800
    Discount: 20%   Selling price: $___

11) Food bill: $56
    Tip:15%   Price: $_____

12) Food bill: $50
    Tipp: 10%   Price: $_____

13) Food bill: $94
    Tip: 25%   Price: $_____

14) Food bill: $48
    Tipp: 30%   Price: $_____

✎ **Find the answer for each word problem.**

15) Nicolas hired a moving company. The company charged $400 for its services, and Nicolas gives the movers a 20% tip. How much does Nicolas tip the movers? $_____

16) Mason has lunch at a restaurant and the cost of his meal is $80. Mason wants to leave a 20% tip. What is Mason's total bill including tip? $_____

17) The sales tax in Texas is 14.45% and an item costs $300. How much is the tax? $_____

18) The price of a table at Best Buy is $520. If the sales tax is 4%, what is the final price of the table including tax? $_____

Percent of Change

Find each percent of change.

1) From 300 to 600. ___ %
2) From 45 ft to 225 ft. ___ %
3) From $60 to $420. ___ %
4) From 30 cm to 120 cm. ___ %
5) From 10 to 30. ___ %
6) From 12 to 30. ___ %
7) From 140 to 210. ___ %
8) From 800 to 400. ___ %
9) From 85 to 51. ___ %
10) From 152 to 76. ___ %

Calculate each percent of change word problem.

11) Bob got a raise, and his hourly wage increased from $32 to $40. What is the percent increase? ___ %

12) The price of a pair of shoes increases from $70 to $112. What is the percent increase? ___ %

13) At a coffee shop, the price of a cup of coffee increased from $1.90 to $2.28. What is the percent increase in the cost of the coffee? ___ %

14) 30 cm are cut from a 120 cm board. What is the percent decrease in length? ___ %

15) In a class, the number of students has been increased from 54 to 81. What is the percent increase? ___ %

16) The price of gasoline rose from $22.4 to $25.76 in one month. By what percent did the gas price rise? ___ %

17) A shirt was originally priced at $19. It went on sale for $22.80. What was the percent that the shirt was discounted? ___ %

Simple Interest

✎ Determine the simple interest for these loans.

1) $210 at 15% for 4 years. $ _____
2) $1,200 at 6% for 3 years. $ _____
3) $950 at 25% for 2 years. $ _____
4) $6,500 at 1.5% for 7 months. $ ____
5) $240 at 5% for 8 months. $ _____
6) $28,000 at 3.5% for 6 years. $ ____
7) $9,600 at 8% for 2 years. $ _____
8) $500 at 4.2% for 5 years. $ _____
9) $700 at 2.8 % for 6 months. $ ____
10) $9,000 at 1.6% for 4 years. $ ____

✎ Calculate each simple interest word problem.

11) A new car, valued at $16,000, depreciates at 3.5% per year. What is the value of the car two year after purchase? $_____

12) Sara puts $9,000 into an investment yielding 8% annual simple interest; she left the money in for three years. How much interest does Sara get at the end of those three years? $_____

13) A bank is offering 12.5% simple interest on a savings account. If you deposit $32,400, how much interest will you earn in one years? $_____

14) $2,400 interest is earned on a principal of $10,000 at a simple interest rate of 12% interest per year. For how many years was the principal invested? _____

15) In how many years will $1,200 yield an interest of $384 at 8% simple interest? _____

16) Jim invested $5,000 in a bond at a yearly rate of 2.5%. He earned $375 in interest. How long was the money invested? _____

Answers of Worksheets – Chapter 3

Simplifying Ratios

1) 3:4
2) 1:10
3) 4:7
4) 1:3
5) 1:10
6) 1:8
7) 2:8
8) 2:5
9) 1:6
10) 7:9
11) 2:3
12) 7:2
13) 10:1
14) 3:4
15) 1:8
16) 5:7
17) 7:9
18) 3:1
19) 1:3
20) 7:1
21) 5:1
22) 5:4
23) 1:2
24) 1:30
25) $\frac{1}{2}$
26) $\frac{3}{5}$
27) $\frac{3}{7}$
28) $\frac{1}{3}$
29) $\frac{1}{3}$
30) $\frac{3}{14}$
31) $\frac{1}{2}$
32) $\frac{1}{5}$
33) $\frac{5}{12}$
34) $\frac{2}{9}$
35) $\frac{11}{13}$
36) $\frac{4}{9}$
37) $\frac{1}{12}$
38) $\frac{3}{11}$
39) $\frac{1}{15}$
40) $\frac{1}{4}$
41) $\frac{4}{3}$
42) $\frac{1}{5}$
43) $\frac{22}{41}$
44) $\frac{1}{9}$
45) $\frac{3}{5}$

Proportional Ratios

1) 12
2) 90
3) 5
4) 24
5) 63
6) 14
7) 23
8) 8
9) 48
10) 50
11) 14.2
12) 18
13) Yes
14) Yes
15) Yes
16) No
17) No
18) No
19) Yes
20) Yes
21) No
22) No
23) Yes
24) Yes
25) 24
26) 392
27) 10
28) 42
29) 48
30) 96
31) 96
32) 68
33) 8.4
34) 8
35) 73.5
36) 94

Similarity and ratios

1) 30
2) 2
3) 10
4) 11
5) 150 feet
6) 8.96 meters
7) 40 feet
8) 187.2 miles

Ratio and Rates Word Problems

1) 4:5
2) 252

3) The ratio for both classes is 9 to 5.

4) Walmart is a better buy.

5) 510, the rate is 30 per hour.

6) $39.20

7) 28

8) 18

Percentage Calculations

1) 1.8

2) 1.6

3) 20

4) 12

5) 27

6) 24.5

7) 4.2

8) 96

9) 43.2

10) 61

11) 182

12) 70.4

13) 88

14) 26.32

15) 25.5

16) 69.36

17) 72

18) 121

19) 50%

20) 80%

21) 5%

22) 25%

23) 12.5%

24) 25%

25) 25%

26) 3%

27) 5%

28) 14.33%

29) 30.30%

30) 54

Percent Problems

1) 50%

2) 40%

3) 160%

4) 4%

5) 1.5%

6) 6%

7) 20%

8) 8%

9) 30%

10) 25%

11) 50%

12) 50%

13) 8%

14) 125%

15) 150%

16) 30%

17) 50%

18) 320%

19) 75%

20) 14.4 ounces

21) 60

22) 75%

23) 20%

Discount, Tax and Tip

1) $486.00

2) $249.60

3) $1,008.00

4) $10,660.00

5) $412.00

6) $370,080

7) $144.00

8) $144.00

9) $56.00

10) $640.00

11) $64.40

12) $55.00

13) $117.50

14) $62.40

15) $80.00

16) $96.00

17) $43.35

18) $540.80

ACT Math Workbook

Percent of Change

1) 100%
2) 400%
3) 600%
4) 300%
5) 200%
6) 150%
7) 50%
8) 50%
9) 40%
10) 50%
11) 25%
12) 60%
13) 20%
14) 25%
15) 50%
16) 15%
17) 20%

Simple Interest

1) $126.00
2) $216.00
3) $475.00
4) $56.875
5) $8.00
6) $5,880.00
7) $1,536.00
8) $105.00
9) $9.80
10) $576.00
11) $14,880.00
12) $2,160.00
13) $4,050.00
14) 2 years
15) 4 years
16) 3 years

Chapter 4:
Exponents and Radicals Expressions

Topics that you will practice in this chapter:

- ✓ Multiplication Property of Exponents
- ✓ Zero and Negative Exponents
- ✓ Division Property of Exponents
- ✓ Powers of Products and Quotients
- ✓ Negative Exponents and Negative Bases
- ✓ Scientific Notation
- ✓ Square Roots
- ✓ Simplifying Radical Expressions
- ✓ Simplifying Radical Expressions Involving Fractions
- ✓ Multiplying Radical Expressions
- ✓ Adding and Subtracting Radical Expressions
- ✓ Domain and Range of Radical Functions
- ✓ Solving Radical Equations

Mathematics is no more computation than typing is literature.
– John Allen Paulos

Multiplication Property of Exponents

✎ Simplify and write the answer in exponential form.

1) $2 \times 2^5 =$

2) $7^2 \times 7 =$

3) $8^3 \times 8^3 =$

4) $9^4 \times 9^3 =$

5) $4^2 \times 4^4 \times 4 =$

6) $5 \times 5^2 \times 5^3 =$

7) $9^3 \times 9^3 \times 9 \times 9 =$

8) $4x \times x =$

9) $x^5 \times x^3 =$

10) $x^6 \times x^2 =$

11) $x^2 \times x^4 \times x^5 =$

12) $7x \times 7x =$

13) $4x^2 \times 5x^3 =$

14) $12x^3 \times x =$

15) $3x^2 \times 3x^2 \times 3x^2 =$

16) $7x^5 \times 2x^3 =$

17) $x^8 \times 2x =$

18) $3x \times 3x^3 =$

19) $6x^2 \times 2x^5 =$

20) $3yx^3 \times 12x =$

21) $8x^3 \times y^5 x^2 =$

22) $4y^7 x^2 \times 3y^2 x^5 =$

23) $9yx^2 \times 4x^5 y^2 =$

24) $10x^4 \times 11x^4 y^4 =$

25) $9x^3 y^4 \times 9x^6 y^2 =$

26) $12x^4 y^4 \times 6xy^3 =$

27) $9xy^4 \times 11x^3 y^3 =$

28) $6x^2 y^4 \times 8x^3 y^6 =$

29) $8x \times y^7 x^2 \times 5y^3 =$

30) $3yx^3 \times 2y^3 x^2 \times 7xy =$

31) $8yx^5 \times 3y^4 x \times 3xy^3 =$

32) $9x^3 \times 11y^4 x^3 \times 2yx^4 =$

Zero and Negative Exponents

✏️ **Evaluate the following expressions.**

1) $1^{-5} =$

2) $2^{-4} =$

3) $2^{-5} =$

4) $3^{0} =$

5) $3^{-2} =$

6) $2^{-7} =$

7) $13^{-2} =$

8) $14^{-2} =$

9) $2^{-8} =$

10) $20^{-2} =$

11) $19^{-1} =$

12) $3^{-6} =$

13) $15^{-2} =$

14) $10^{-2} =$

15) $16^{-2} =$

16) $30^{-2} =$

17) $8^{-4} =$

18) $3^{-7} =$

19) $2^{-10} =$

20) $10^{-3} =$

21) $18^{-2} =$

22) $25^{-2} =$

23) $40^{-2} =$

24) $50^{-2} =$

25) $11^{-3} =$

26) $22^{-2} =$

27) $17^{-2} =$

28) $3^{-8} =$

29) $4^{-5} =$

30) $60^{-2} =$

31) $\left(\frac{1}{3}\right)^{-2} =$

32) $\left(\frac{1}{5}\right)^{-3} =$

33) $\left(\frac{1}{8}\right)^{-2} =$

34) $\left(\frac{2}{5}\right)^{-2} =$

35) $\left(\frac{1}{15}\right)^{-2} =$

36) $\left(\frac{7}{12}\right)^{-2} =$

37) $\left(\frac{1}{20}\right)^{-2} =$

38) $\left(\frac{1}{7}\right)^{-3} =$

39) $\left(\frac{2}{3}\right)^{-5} =$

40) $\left(\frac{9}{11}\right)^{-1} =$

41) $\left(\frac{8}{9}\right)^{-2} =$

42) $\left(\frac{1}{8}\right)^{-3} =$

Division Property of Exponents

✎ Simplify.

1) $\dfrac{5^2}{5^6} =$

2) $\dfrac{6^9}{6^5} =$

3) $\dfrac{9^7}{9} =$

4) $\dfrac{3}{3^3} =$

5) $\dfrac{2x}{x^8} =$

6) $\dfrac{4 \times 4^5}{4^5 \times 4^2} =$

7) $\dfrac{12^6}{12^2} =$

8) $\dfrac{7 \times 7^9}{7^2 \times 7^4} =$

9) $\dfrac{4^5 \times 4^8}{4^2 \times 4^{11}} =$

10) $\dfrac{20x}{40x^4} =$

11) $\dfrac{8x^9}{9x^6} =$

12) $\dfrac{24x^3}{16x^5} =$

13) $\dfrac{25x^2}{50y^8} =$

14) $\dfrac{60xy^5}{12x^4y^2} =$

15) $\dfrac{8x^7}{12x} =$

16) $\dfrac{48x^2y^4}{16x^5} =$

17) $\dfrac{50x^6}{25x^9y^{14}} =$

18) $\dfrac{90yx^8}{15yx^9} =$

19) $\dfrac{18x^9y}{36x^{12}y^3} =$

20) $\dfrac{9x^8}{81x^8} =$

21) $\dfrac{9x^{-7}}{11x^{-3}} =$

Powers of Products and Quotients

✎ **Simplify.**

1) $(4^2)^3 =$

2) $(5^2)^2 =$

3) $(3 \times 3^2)^3 =$

4) $(3 \times 2^3)^2 =$

5) $(15^2 \times 15^2)^5 =$

6) $(7^2 \times 7^3)^4 =$

7) $(9 \times 9^2)^2 =$

8) $(4^6)^3 =$

9) $(7x^7)^3 =$

10) $(8x^4y^3)^2 =$

11) $(3x^3y^2)^4 =$

12) $(4x^2y^2)^2 =$

13) $(3x^5y^2)^3 =$

14) $(4x^3y^2)^3 =$

15) $(2x^3x)^5 =$

16) $(6x^4x^2)^2 =$

17) $(7x^{12}y^5)^2 =$

18) $(5x^7x^4)^3 =$

19) $(8x^2 \times 6x)^2 =$

20) $(9x^{14}y^3)^3 =$

21) $(5x^4y^2)^4 =$

22) $(3x^3y^7)^5 =$

23) $(8x \times 2y^3)^2 =$

24) $\left(\frac{8x}{x^3}\right)^3 =$

25) $\left(\frac{x^4y^5}{x^3y^5}\right)^7 =$

26) $\left(\frac{36xy}{6x^5}\right)^2 =$

27) $\left(\frac{x^4}{x^5y^2}\right)^3 =$

28) $\left(\frac{xy^2}{x^3y^8}\right)^{-3} =$

29) $\left(\frac{5xy^7}{x^2}\right)^3 =$

30) $\left(\frac{xy^5}{2xy^3}\right)^{-6} =$

Negative Exponents and Negative Bases

Simplify.

1) $-4^{-2} =$

2) $-7^{-1} =$

3) $-5^{-2} =$

4) $-x^{-9} =$

5) $10x^{-2} =$

6) $-7x^{-4} =$

7) $-15x^{-4} =$

8) $-15x^{-7}y^{-4} =$

9) $32x^{-9}y^{-3} =$

10) $45a^{-7}b^{-3} =$

11) $-25x^3y^{-5} =$

12) $-\dfrac{18}{x^{-9}} =$

13) $-\dfrac{13x}{a^{-8}} =$

14) $\left(-\dfrac{1}{3}\right)^{-4} =$

15) $\left(-\dfrac{3}{4}\right)^{-3} =$

16) $-\dfrac{12}{a^{-6}b^{-4}} =$

17) $-\dfrac{48x}{x^{-6}} =$

18) $-\dfrac{a^{-12}}{b^{-5}} =$

19) $-\dfrac{27}{x^{-5}} =$

20) $\dfrac{12b}{-48c^{-6}} =$

21) $\dfrac{24ab}{a^{-4}b^{-3}} =$

22) $-\dfrac{8n^{-7}}{40p^{-9}} =$

23) $\dfrac{9ab^{-6}}{-5c^{-2}} =$

24) $\left(\dfrac{2a}{3c}\right)^{-4} =$

25) $\left(-\dfrac{8x}{5yz}\right)^{-2} =$

26) $\dfrac{9ab^{-6}}{-4c^{-3}} =$

27) $\left(-\dfrac{x^3}{x^4}\right)^{-5} =$

28) $\left(-\dfrac{x^{-3}}{3x^3}\right)^{-3} =$

29) $\left(-\dfrac{x^{-6}}{x^4}\right)^{-3} =$

Scientific Notation

✎ Write each number in scientific notation.

1) $0.226 =$

2) $0.05 =$

3) $4.8 =$

4) $90 =$

5) $120 =$

6) $0.123 =$

7) $82 =$

8) $5,400 =$

9) $2,460 =$

10) $75,300 =$

11) $61,000,000 =$

12) $0.00009 =$

13) $468,000 =$

14) $0.00458 =$

15) $0.000087 =$

16) $31,800,000 =$

17) $950,000 =$

18) $9,000,000,000 =$

19) $0.0007 =$

20) $0.00041 =$

✎ Write each number in standard notation.

21) $4 \times 10^{-2} =$

22) $7 \times 10^{-4} =$

23) $4.3 \times 10^{6} =$

24) $7 \times 10^{-4} =$

25) $8.7 \times 10^{-3} =$

26) $12 \times 10^{5} =$

27) $35 \times 10^{3} =$

28) $1.89 \times 10^{5} =$

29) $13 \times 10^{-6} =$

30) $7.3 \times 10^{-4} =$

Square Roots

✎ **Find the value each square root.**

1) $\sqrt{64}$ = ___

2) $\sqrt{4}$ = ___

3) $\sqrt{289}$ = ___

4) $\sqrt{0.25}$ = ___

5) $\sqrt{0.01}$ = ___

6) $\sqrt{0.09}$ = ___

7) $\sqrt{1,600}$ = ___

8) $\sqrt{2.25}$ = ___

9) $\sqrt{0}$ = ___

10) $\sqrt{0.04}$ = ___

11) $\sqrt{0.36}$ = ___

12) $\sqrt{0.81}$ = ___

13) $\sqrt{0.49}$ = ___

14) $\sqrt{1.21}$ = ___

15) $\sqrt{1.69}$ = ___

16) $\sqrt{0.16}$ = ___

17) $\sqrt{529}$ = ___

18) $\sqrt{625}$ = ___

19) $\sqrt{0.81}$ = ___

20) $\sqrt{20}$ = ___

21) $\sqrt{50}$ = ___

22) $\sqrt{676}$ = ___

23) $\sqrt{270}$ = ___

24) $\sqrt{32}$ = ___

✎ **Evaluate.**

25) $\sqrt{4} \times \sqrt{16}$ = _____

26) $\sqrt{49} \times \sqrt{64}$ = _____

27) $\sqrt{2} \times \sqrt{8}$ = _____

28) $\sqrt{17} \times \sqrt{17}$ = _____

29) $\sqrt{13} \times \sqrt{13}$ = _____

30) $\sqrt{15} \times \sqrt{15}$ = _____

31) $\sqrt{19} + \sqrt{19}$ = _____

32) $\sqrt{1} + \sqrt{1}$ = _____

33) $8\sqrt{7} - 2\sqrt{7}$ = _____

34) $7\sqrt{10} \times 6\sqrt{10}$ = _____

35) $9\sqrt{5} \times 2\sqrt{5}$ = _____

36) $8\sqrt{3} - \sqrt{12}$ = _____

Simplifying Radical Expressions

✏️ **Simplify.**

1) $\sqrt{13y^2} =$

2) $\sqrt{60x^3} =$

3) $\sqrt[3]{27a} =$

4) $\sqrt{81x^2} =$

5) $\sqrt{150a} =$

6) $\sqrt[3]{135w^3} =$

7) $\sqrt{200x} =$

8) $\sqrt{192v} =$

9) $\sqrt[3]{64x} =$

10) $\sqrt{84x^3} =$

11) $\sqrt{121x^2} =$

12) $\sqrt[3]{48a} =$

13) $\sqrt{480} =$

14) $\sqrt{1,575p^2} =$

15) $\sqrt{108m^6} =$

16) $\sqrt{198x^3y^2} =$

17) $\sqrt{169x^2y^3} =$

18) $\sqrt{25a^6} =$

19) $\sqrt{50x^2y^3} =$

20) $\sqrt[3]{512y^3} =$

21) $2\sqrt{144x^2} =$

22) $3\sqrt{400x^2} =$

23) $\sqrt[3]{189xy^4} =$

24) $\sqrt[3]{1,331x^3y^5} =$

25) $3\sqrt{150a} =$

26) $\sqrt[3]{729y} =$

27) $3\sqrt{18xyr^3} =$

28) $6\sqrt{225x^2yz^6} =$

29) $3\sqrt[3]{125x^3y^2} =$

30) $7\sqrt{12a^2bc^4} =$

31) $4\sqrt[3]{1,000x^9y^{15}} =$

Multiplying Radical Expressions

✎ **Simplify.**

1) $\sqrt{11} \times \sqrt{11} =$

2) $\sqrt{5} \times \sqrt{15} =$

3) $\sqrt{3} \times \sqrt{12} =$

4) $\sqrt{20} \times \sqrt{25} =$

5) $\sqrt{5} \times (-2)\sqrt{35} =$

6) $2\sqrt{12} \times \sqrt{3} =$

7) $4\sqrt{24} \times \sqrt{6} =$

8) $\sqrt{5} \times (-\sqrt{75}) =$

9) $\sqrt{88} \times \sqrt{40} =$

10) $2\sqrt{45} \times 4\sqrt{105} =$

11) $\sqrt{32}(2 + \sqrt{2}) =$

12) $\sqrt{13x^2} \times \sqrt{13x} =$

13) $-7\sqrt{12} \times \sqrt{3} =$

14) $5\sqrt{19x^3} \times \sqrt{19x^3} =$

15) $\sqrt{15x^2} \times \sqrt{5x} =$

16) $-8\sqrt{2x} \times \sqrt{6x^5} =$

17) $-4\sqrt{5x} \times 5\sqrt{45x^2} =$

18) $-3\sqrt{27}(3 + \sqrt{15}) =$

19) $\sqrt{8x}(3 - \sqrt{2x}) =$

20) $\sqrt{5x}(10\sqrt{5x} + \sqrt{40}) =$

21) $\sqrt{18r}(6 + \sqrt{6}) =$

22) $-12\sqrt{3x} \times 3\sqrt{15x^3} =$

23) $-4\sqrt{27x} \times 6\sqrt{3x}$

24) $-3\sqrt{10v^2}(-3\sqrt{15v}) =$

25) $(\sqrt{8} - 3)(\sqrt{8} + \sqrt{9}) =$

26) $(-3\sqrt{5} + 7)(\sqrt{5} - 3) =$

27) $(3 - 4\sqrt{5})(-2 + \sqrt{4}) =$

28) $(13 - 2\sqrt{5})(3 - \sqrt{5}) =$

29) $(5 - \sqrt{3x})(5 + \sqrt{3x}) =$

30) $(-6 + 3\sqrt{3r})(-6 + \sqrt{3r}) =$

31) $(-\sqrt{5n} + 8)(-\sqrt{5} - 8) =$

32) $(-3 + 3\sqrt{2})(3 - 2\sqrt{2x}) =$

WWW.MathNotion.Com

Simplifying Radical Expressions Involving Fractions

✎ **Simplify.**

1) $\frac{\sqrt{3}}{\sqrt{2}} =$

2) $\frac{\sqrt{24}}{\sqrt{40}} =$

3) $\frac{\sqrt{12}}{2\sqrt{6}} =$

4) $\frac{21}{\sqrt{5}} =$

5) $\frac{15\sqrt{8r}}{\sqrt{m^5}} =$

6) $\frac{8\sqrt{2}}{\sqrt{m}} =$

7) $\frac{15\sqrt{25n^2}}{5\sqrt{15n}} =$

8) $\frac{\sqrt{8x^3y^5}}{\sqrt{2y^2x^4}} =$

9) $\frac{2}{2+\sqrt{5}} =$

10) $\frac{2-12\sqrt{x}}{\sqrt{24x}} =$

11) $\frac{2\sqrt{x}}{\sqrt{x}-\sqrt{y}} =$

12) $\frac{3-\sqrt{5}}{5-\sqrt{3}} =$

13) $\frac{5+\sqrt{12}}{5-\sqrt{3}} =$

14) $\frac{8}{-3-3\sqrt{3}} =$

15) $\frac{5}{2+\sqrt{15}} =$

16) $\frac{\sqrt{11}-\sqrt{7}}{\sqrt{7}-\sqrt{11}} =$

17) $\frac{\sqrt{8}+\sqrt{2}}{\sqrt{2}-\sqrt{8}} =$

18) $\frac{2\sqrt{2}-\sqrt{3}}{3\sqrt{2}+\sqrt{3}} =$

19) $\frac{\sqrt{11}+3\sqrt{5}}{3-\sqrt{11}} =$

20) $\frac{\sqrt{7}+\sqrt{13}}{13-\sqrt{7}} =$

21) $\frac{\sqrt{125a^7b^5}}{\sqrt{5ab^4}} =$

22) $\frac{72\sqrt{24m^3}}{9\sqrt{m}} =$

Adding and Subtracting Radical Expressions

✏️ **Simplify.**

1) $\sqrt{5} + \sqrt{20} =$

2) $7\sqrt{44} + 7\sqrt{11} =$

3) $9\sqrt{3} - 3\sqrt{12} =$

4) $8\sqrt{9} - 5\sqrt{3} =$

5) $9\sqrt{80} - 9\sqrt{20} =$

6) $-\sqrt{32} - 5\sqrt{8} =$

7) $-12\sqrt{16} - 9\sqrt{64} =$

8) $15\sqrt{8} + 6\sqrt{32} =$

9) $16\sqrt{9} - 12\sqrt{36} =$

10) $-7\sqrt{7} + 11\sqrt{63} =$

11) $-24\sqrt{13} + 18\sqrt{117} =$

12) $25\sqrt{5} - 12\sqrt{45} =$

13) $-8\sqrt{99} + 2\sqrt{11} =$

14) $6\sqrt{3} - 2\sqrt{12} =$

15) $8\sqrt{20} + 3\sqrt{5} =$

16) $5\sqrt{28} - 8\sqrt{63} =$

17) $\sqrt{144} - \sqrt{121} =$

18) $6\sqrt{18} - 2\sqrt{2} =$

19) $-12\sqrt{7} + 21\sqrt{28} =$

20) $5\sqrt{60} - 5\sqrt{15} =$

21) $6\sqrt{54} - 3\sqrt{6} =$

22) $-4\sqrt{3} + 8\sqrt{75} =$

23) $-9\sqrt{20} - 7\sqrt{5} =$

24) $-\sqrt{216x} + 6\sqrt{6x} =$

25) $\sqrt{14y^2} + y\sqrt{126} =$

26) $\sqrt{11mn^2} + n\sqrt{99m} =$

27) $-8\sqrt{48a} - 2\sqrt{3a} =$

28) $-15\sqrt{17ab} - 10\sqrt{68ab} =$

29) $\sqrt{92x^2y} + x\sqrt{23y} =$

30) $5\sqrt{5a} + 4\sqrt{80a} =$

Answers of Worksheets – Chapter 4

Multiplication Property of Exponents

1) 2^6
2) 7^3
3) 8^6
4) 9^7
5) 4^7
6) 5^6
7) 9^8
8) $4x^2$
9) x^8
10) x^8
11) x^{11}
12) $49x^2$
13) $20x^5$
14) $12x^4$
15) $27x^6$
16) $14x^8$
17) $2x^8$
18) $9x^4$
19) $12x^7$
20) $36x^4 y$
21) $8x^5 y^5$
22) $12x^7 y^9$
23) $36x^7 y^3$
24) $110x^8 y^4$
25) $81x^9 y^6$
26) $72x^5 y^7$
27) $99x^4 y^7$
28) $48x^5 y^{10}$
29) $40x^3 y^{10}$
30) $42x^6 y^5$
31) $72x^7 y^8$
32) $198x^{10} y^5$

Zero and Negative Exponents

1) 1
2) $\frac{1}{16}$
3) $\frac{1}{32}$
4)
5) $\frac{1}{9}$
6) $\frac{1}{128}$
7) $\frac{1}{169}$
8) $\frac{1}{196}$
9) $\frac{1}{256}$
10) $\frac{1}{400}$
11) $\frac{1}{19}$
12) $\frac{1}{729}$
13) $\frac{1}{225}$
14) $\frac{1}{100}$
15) $\frac{1}{256}$
16) $\frac{1}{900}$
17) $\frac{1}{4,096}$
18) $\frac{1}{2,187}$
19) $\frac{1}{1,024}$
20) $\frac{1}{1,000}$
21) $\frac{1}{324}$
22) $\frac{1}{625}$
23) $\frac{1}{1,600}$
24) $\frac{1}{2,500}$
25) $\frac{1}{1,331}$
26) $\frac{1}{484}$
27) $\frac{1}{289}$
28) $\frac{1}{6,561}$
29) $\frac{1}{1,024}$
30) $\frac{1}{3,600}$
31)
32)
33)
34) 6.25
35) 225
36) $\frac{144}{49}$
37) 400
38) 343
39) $\frac{243}{32}$
40) $\frac{11}{9}$
41) $\frac{81}{64}$
42) 512

Division Property of Exponents

1) $\frac{1}{5^4}$
2) 6^4
3) 9^6
4) $\frac{1}{3^2}$
5) $\frac{2}{x^7}$
6) $\frac{1}{4}$
7) 12^4
8) 7^4
9) 1
10) $\frac{1}{2x^3}$
11) $\frac{8x^3}{9}$
12) $\frac{3}{2x^2}$
13) $\frac{x^2}{2y^8}$
14) $\frac{5y^3}{x^3}$
15) $\frac{2x^6}{3}$

16) $\frac{3y^4}{x^3}$ 18) $\frac{6}{x}$ 19) $\frac{1}{2x^3y^2}$ 21) $\frac{9}{11x^4}$

17) $\frac{2}{x^3y^{14}}$ 20) $\frac{1}{9}$

Powers of Products and Quotients

1) 4^6
2) 5^4
3) 3^9
4) 24^2
5) 15^{20}
6) 7^{20}
7) 9^6
8) 4^{18}
9) $343x^{21}$
10) $64x^8y^6$
11) $81x^{12}y^8$
12) $16x^4y^4$
13) $27x^{15}y^6$
14) $64x^9y^6$
15) $32x^{20}$
16) $36x^{12}$
17) $49x^{24}y^{10}$
18) $125x^{33}$
19) $2,304x^6$
20) $729x^{42}y^9$
21) $625x^{16}y^9$
22) $243x^{15}y^8$
23) $256x^2y^6$
24) $\frac{512}{x^6}$
25) x^7
26) $\frac{36y^2}{x^8}$
27) $\frac{1}{x^3y^6}$
28) x^6y^{18}
29) $\frac{125y^{21}}{x^3}$
30) $\frac{64}{y^{12}}$

Negative Exponents and Negative Bases

1) $-\frac{1}{16}$
2) $-\frac{1}{7}$
3) $-\frac{1}{25}$
4) $-\frac{1}{x^9}$
5) $\frac{10}{x^2}$
6) $-\frac{7}{x^4}$
7) $-\frac{15}{x^4}$
8) $-\frac{15}{x^7y^4}$
9) $\frac{32}{x^9y^3}$
10) $\frac{45}{a^7b^3}$
11) $-\frac{25x^3}{y^5}$
12) $-18x^9$
13) $-13xa^8$
14) 81
15) $-\frac{64}{27}$
16) $-12a^6b^4$
17) $-48x^7$
18) $-\frac{b^5}{a^{12}}$
19) $-27x^5$
20) $-\frac{bc^6}{4}$
21) $24a^5b^4$
22) $-\frac{p^9}{5n^7}$
23) $-\frac{9ac^2}{5b^6}$
24) $\frac{81c^4}{16a^4}$
25) $\frac{25y^2z^2}{64x^2}$
26) $-\frac{9ac^3}{4b^6}$
27) $-x^5$
28) $-27x^{18}$
29) $-x^{30}$

Writing Scientific Notation

1) 2.26×10^{-1}
2) 5×10^{-2}
3) 4.8×10^0

ACT Math Workbook

4) 9×10^1
5) 1.2×10^2
6) 1.23×10^{-1}
7) 8.2×10^1
8) 5.4×10^3
9) 2.46×10^3
10) 7.53×10^4
11) 61×10^6
12) 9×10^{-5}
13) 4.68×10^5
14) 4.58×10^{-3}
15) 8.7×10^{-5}
16) 3.18×10^7
17) 9.5×10^5
18) 9×10^9
19) 7×10^{-4}
20) 4.1×10^{-4}
21) 0.04
22) 0.0007
23) 4,300,000
24) 0.0007
25) 0.0087
26) 1,200,000
27) 35,000
28) 189,000
29) 0.000013
30) 0.00073

Square Roots

1) 8
2) 2
3) 17
4) 0.5
5) 0.1
6) 0.3
7) 40
8) 1.5
9) 0
10) 0.2
11) 0.6
12) 0.9
13) 0.7
14) 1.1
15) 1.3
16) 0.4
17) 23
18) 25
19) 0.9
20) $2\sqrt{5}$
21) $5\sqrt{2}$
22) 26
23) $3\sqrt{30}$
24) $4\sqrt{2}$
25) 8
26) 56
27) 4
28) 17
29) 13
30) 15
31) $2\sqrt{19}$
32) 2
33) $6\sqrt{7}$
34) 420
35) 90
36) $6\sqrt{3}$

Simplifying radical expressions

1) $y\sqrt{13}$
2) $2x\sqrt{15x}$
3) $3\sqrt[3]{a}$
4) $9x$
5) $5\sqrt{6a}$
6) $3w\sqrt[3]{5}$
7) $10\sqrt{2x}$
8) $8\sqrt{3v}$
9) $4\sqrt[3]{x}$
10) $2x\sqrt{21x}$
11) $11x$
12) $2\sqrt[3]{6a}$
13) $4\sqrt{30}$
14) $15p\sqrt{7}$
15) $6m^3\sqrt{3}$
16) $3x.y\sqrt{22x}$
17) $13xy\sqrt{y}$
18) $5a^3$
19) $5xy\sqrt{2y}$
20) $8y$
21) $24x$
22) $60x$
23) $3y\sqrt[3]{7xy}$
24) $11xy\sqrt[3]{y^2}$
25) $15\sqrt{6a}$
26) $9\sqrt[3]{y}$

27) $9r\sqrt{2xyr}$

28) $90xz^3\sqrt{y}$

29) $15x\sqrt[3]{y^2}$

30) $14ac^2\sqrt{b}$

31) $40x^3y^{15}$

Multiplying radical expressions

1) 11

2) $5\sqrt{3}$

3) 6

4) $10\sqrt{5}$

5) $-10\sqrt{7}$

6) 12

7) 48

8) $-5\sqrt{15}$

9) $8\sqrt{55}$

10) $120\sqrt{21}$

11) $8\sqrt{2}+8$

12) $13x\sqrt{x}$

13) -42

14) $95x^3$

15) $5x\sqrt{3x}$

16) $-16x^3\sqrt{3}$

17) $-300x\sqrt{x}$

18) $-27\sqrt{3}-27\sqrt{5}$

19) $6\sqrt{2x}-4x$

20) $50x+2\sqrt{50x}$

21) $18\sqrt{2r}+6\sqrt{3r}$

22) $-108x^2\sqrt{5}$

23) $-216x$

24) $47v\sqrt{6v}$

25) -1

26) $16\sqrt{5}-36$

27) 0

28) $49-19\sqrt{5}$

29) $28-3x$

30) $9r-24\sqrt{3r}+36$

31) $5\sqrt{n}-64$

32) $-9+6\sqrt{2x}+9\sqrt{2}-12\sqrt{x}$

Simplifying radical expressions involving fractions

1) $\frac{\sqrt{6}}{2}$

2) $\frac{\sqrt{15}}{5}$

3) $\frac{\sqrt{2}}{2}$

4) $\frac{21\sqrt{5}}{5}$

5) $\frac{30\sqrt{2mr}}{m^3}$

6) $\frac{8\sqrt{2m}}{m}$

7) $\sqrt{15n}$

8) $\frac{2y\sqrt{xy}}{x}$

9) $-2(2-\sqrt{2})$

10) $\frac{\sqrt{6x}\,(1-6\sqrt{x})}{6x}$

11) $\frac{2\sqrt{x}\,(\sqrt{x}+\sqrt{y})}{x-y}$

12) $\frac{15+3\sqrt{3}-5\sqrt{5}-\sqrt{15}}{22}$

13) $\frac{31+15\sqrt{3}}{22}$

14) $-\frac{4(\sqrt{3}-1)}{3}$

15) $-\frac{5(2-\sqrt{15})}{11}$

16) -1

17) -3

18) $\frac{3-\sqrt{6}}{3}$

19) $-\frac{3\sqrt{11}+11+9\sqrt{5}+3\sqrt{55}}{2}$

20) $\frac{13\sqrt{7}+7+13\sqrt{13}+\sqrt{91}}{162}$

21) $5a^3\sqrt{b}$

22) $16\sqrt{6}\,m$

Adding and subtracting radical expressions

1) $2\sqrt{5}$
2) $21\sqrt{11}$
3) $-5\sqrt{3}$
4) $24-5\sqrt{3}$
5) $18\sqrt{5}$
6) $-14\sqrt{2}$
7) -120
8) $54\sqrt{2}$
9) -24
10) $26\sqrt{7}$
11) $30\sqrt{13}$
12) $-11\sqrt{5}$
13) $-22\sqrt{11}$
14) $2\sqrt{3}$
15) $19\sqrt{5}$
16) $-14\sqrt{7}$
17) 1
18) $16\sqrt{2}$
19) $30\sqrt{7}$
20) $5\sqrt{15}$
21) $15\sqrt{6}$
22) $36\sqrt{3}$
23) $-25\sqrt{5}$
24) 0
25) $4y\sqrt{14}$
26) $4n\sqrt{11m}$
27) $-34\sqrt{3a}$
28) $-35\sqrt{17ab}$
29) $-x\sqrt{23y}$
30) $21\sqrt{5a}$

Chapter 5:
Algebraic Expressions

Topics that you will practice in this chapter:

- ✓ Simplifying Variable Expressions
- ✓ Simplifying Polynomial Expressions
- ✓ Translate Phrases into an Algebraic Statement
- ✓ The Distributive Property
- ✓ Evaluating One Variable Expressions
- ✓ Evaluating Two Variables Expressions
- ✓ Combining like Terms

Mathematics is, as it were, a sensuous logic, and relates to philosophy as do the arts, music, and plastic art to poetry. — *K. Shegel*

Simplifying Variable Expressions

✎ **Simplify each expression.**

1) $3(x + 8) =$

2) $(-4)(7x - 3) =$

3) $11x + 8 - 7x =$

4) $-6 - 2x^2 - 9x^2 =$

5) $8 + 17x^2 + 6 =$

6) $9x^2 + 13x + 19x^2 =$

7) $7x^2 - 15x^2 + 3x =$

8) $8x^2 - 11x - 3x =$

9) $3x + 9(1 - 4x) =$

10) $14x + 2(20x - 4) =$

11) $6(-3x - 7) - 22 =$

12) $7x^2 + (-12x) =$

13) $x - 8 + 15 - 7x =$

14) $3 - 6x + 12 - 3x =$

15) $20x - 14 + 27 + 12x =$

16) $(-7)(6x - 5) + 14x =$

17) $11x - 4(3 - 7x) =$

18) $22x + 3(5x + 2) + 11 =$

19) $4(-3x + 8) + 10x =$

20) $16x - 3x(2x + 7) =$

21) $9x + 12x(2 - 4x) =$

22) $5x(-4x + 11) + 18x =$

23) $20x + 24x + 3x^2 =$

24) $7x(x - 7) - 28 =$

25) $7x - 12 + 5x + 3x^2 =$

26) $4x^2 - 9x - 12x =$

27) $8x - 22x^2 - 21x^2 - 11 =$

28) $9 + 3x^2 - 8x^2 - 27x =$

29) $14x + 2x^2 + 4x + 28 =$

30) $7x^2 + 45x + 12x^2 =$

31) $25 + 15x^2 + 9x - 3x^2 =$

32) $17x - 32x - 2x^2 + 30 =$

Simplifying Polynomial Expressions

✏️ **Simplify each polynomial.**

1) $(5x^4 + 2x^2) - (11x + 6x^2) = $ _____

2) $(x^7 + 6x^4) - (8x^4 + 4x^2) = $ _____

3) $(24x^5 + 8x^3) - (x^3 - 12x^5) = $ _____

4) $14x - 9x^5 - 4(7x^5 + 7x^3) = $ _____

5) $(7x^4 - 5) + 2(4x^2 - 8x^4) = $ _____

6) $(9x^5 - 3x) - 3(8x^5 - 3x^4) = $ _____

7) $4(2x - 4x^4) - 5(3x^4 + x^2) = $ _____

8) $(4x^2 - 2x) - (5x^3 + 9x^2) = $ _____

9) $8x^4 - (9x^6 + 2x) + 2x^2 = $ _____

10) $x^5 - 3(x^3 + 2x) + 9x = $ _____

11) $(4x^2 - 2x^5) - (4x^5 - 2x^2) = $ _____

12) $8x^3 - 8x^5 + 17x^4 - 12x^5 = $ _____

13) $4x^3 - 9x^7 + 18x^7 - 24x^6 = $ _____

14) $4x^5 + 13x^3 - 17x^5 + 24x = $ _____

15) $7x^6 - 9x^7 + 5x^6 - 12x^3 = $ _____

16) $4x^4 + 19x - 3x^3 - 21x^4 = $ _____

Translate Phrases into an Algebraic Statement

✏️ **Write an algebraic expression for each phrase.**

1) 13 multiplied by x. _____

2) Subtract 15 from y. _____

3) 22 divided by x. _____

4) 27 decreased by y. _____

5) Add y to 31. _____

6) The square of 7. _____

7) x raised to the seventh power. _____

8) The sum of five and a number. _____

9) The difference between forty-nine and y. _____

10) The quotient of eight and a number. _____

11) The quotient of the square of x and 34. _____

12) The difference between x and 14 is 41. _____

13) 7 times b reduced by the square of a. _____

14) Subtract the product of a and b from 51. _____

The Distributive Property

✏️ Use the distributive property to simply each expression.

1) $4(2 + 5x) =$

2) $5(2 + 4x) =$

3) $6(5x - 5) =$

4) $(6x - 3)(-7) =$

5) $(-4)(x + 8) =$

6) $(4 + 4x)6 =$

7) $(-5)(8 - 7x) =$

8) $-(-3 - 12x) =$

9) $(-8x + 3)(-5) =$

10) $(-5)(x - 11) =$

11) $-(8 - 2x) =$

12) $3(7 + 4x) =$

13) $4(8 + 3x) =$

14) $(-8x + 2)5 =$

15) $(4 - 7x)(-9) =$

16) $(-12)(3x + 5) =$

17) $(9 - 3x)5 =$

18) $4(4 + 7x) =$

19) $12(3x - 6) =$

20) $(-7x + 5)4 =$

21) $(4 - 9x)(-2) =$

22) $(-15)(2x - 3) =$

23) $(14 - 3x)3 =$

24) $(-5)(10x - 4) =$

25) $(5 - 7x)(-12) =$

26) $(-8)(2x + 9) =$

27) $(-5 + 8x)(-7) =$

28) $(-6)(2 - 15x) =$

29) $13(4x - 6) =$

30) $(-15x + 13)(-4) =$

31) $(-9)(3x - 2) + 2(x + 5) =$

32) $(-9)(2x + 2) - (7 + 4x) =$

Evaluating One Variable Expressions

✎ **Evaluate each expression using the value given.**

1) $8 - x$, $x = 5$

2) $x - 10$, $x = 6$

3) $3x - 6$, $x = 5$

4) $x - 15$, $x = -2$

5) $12 - x$, $x = 4$

6) $x + 7$, $x = 1$

7) $2x + 9$, $x = 7$

8) $x + (-4)$, $x = -7$

9) $2x + 9$, $x = 4$

10) $3x + 10$, $x = -2$

11) $18 + 2x - 4$, $x = -1$

12) $18 - 6x$, $x = 2$

13) $8x - 2$, $x = 4$

14) $2x - 17$, $x = 8$

15) $13x - 12$, $x = 3$

16) $8 - 5x$, $x = -2$

17) $3(5x + 4)$, $x = 5$

18) $4(-2x - 7)$, $x = 3$

19) $7x - 5x + 12$, $x = 2$

20) $(8x + 4) \div 2$, $x = 6$

21) $(x + 15) \div 4$, $x = 9$

22) $6x - 10 + 3x$, $x = -5$

23) $(7 - 4x)(-3)$, $x = -3$

24) $12x^2 + 5x - 4$, $x = 2$

25) $x^2 - 15x$, $x = -4$

26) $3x(3 - 6x)$, $x = 2$

27) $13x + 8 - 6x^2$, $x = -3$

28) $(-2)(15x - 11 + 4x)$, $x = 4$

29) $(-6) + \frac{x}{6} + x$, $x = 18$

30) $(-9) + \frac{x}{4}$, $x = 32$

31) $\left(-\frac{45}{x}\right) - 5 + 2x$, $x = 9$

32) $\left(-\frac{36}{x}\right) - 9 + 3x$, $x = 3$

Evaluating Two Variables Expressions

✎ **Evaluate each expression using the values given.**

1) $5x - y$,

   $x = 4, y = 3$

2) $3x + 2y$,

   $x = -2, y = 2$

3) $-6a + 5b$,

   $a = 3, b = 1$

4) $3x + 7 - y$,

   $x = 8, y = 4$

5) $5z + 12 - 3k$,

   $z = 5, k = 2$

6) $6(-x - 3y)$,

   $x = 5, y = 4$

7) $7a + 4b$,

   $a = 3, b = 5$

8) $8x \div 4y$,

   $x = 6, y = 4$

9) $2x + 18 + 4y$,

   $x = -3, y = 3$

10) $5a - (18 - 2b)$,

    $a = 5, b = 8$

11) $6z + 12 + 3k$,

    $z = -3, k = 3$

12) $2xy + 6 + 7x$,

    $x = 5, y = 3$

13) $6x + 2y - 9 + 3$,

    $x = 3, y = 2$

14) $\left(-\frac{21}{x}\right) + 6 + 3y$,

    $x = 7, y = 4$

15) $(-4)(-3a - b)$,

    $a = 2, b = 6$

16) $18 + 4x + 9 - 5y$,

    $x = 6, y = 4$

17) $7x + 5 - 6y + 11$,

    $x = 9, y = 3$

18) $9 + 4(-5x - 3y)$,

    $x = 4, y = 5$

19) $3x + 15 + 6y$,

    $x = 2, y = 4$

20) $7a - (4a - 2b) + 8$,

    $a = 3, b = 1$

Combining like Terms

✏️ **Simplify each expression.**

1) $7x + 2x + 8 =$

2) $3(6x - 2) =$

3) $10x - 12x + 8 =$

4) $20x - 32x + 14 =$

5) $16x - 6x - 12 =$

6) $18x - 21 + 4x =$

7) $15 - (3x + 9) =$

8) $-14x + 7 - 11x =$

9) $5x - 10 - 3x + 1 =$

10) $24x + 7x - 22 =$

11) $14x + 8x - 2 =$

12) $(-4x + 2)8 =$

13) $34 + 6x + 8x - 4 =$

14) $3(x - 8x) - 5 =$

15) $4(2x + 7) + 5x =$

16) $x - 27 - 9x =$

17) $3(5 + 4x) - 8x =$

18) $41x + 24 + 3x =$

19) $(-8x) + 30 + 15x =$

20) $(-4x) - 12 + 19x =$

21) $5(2x + 6) + 9x =$

22) $3(6 - 7x) - 11x =$

23) $-8x - (16 - 14x) =$

24) $(-9) - (6)(5x + 9) =$

25) $(-4)(6x - 5) - 12x =$

26) $-34x + 14 + 9x - 21x =$

27) $5(-13x + 6) - 24x =$

28) $-7x - 20 + 15x =$

29) $42x - 31x + 15 - 12x =$

30) $4(8x + 5x) - 17 =$

31) $54 - 22x - 28 - 19x =$

32) $-9(-7x - 11x) + 58x =$

Answers of Worksheets – Chapter 5

Simplifying Variable Expressions

1) $3x + 24$
2) $-28x + 12$
3) $4x + 8$
4) $-11x^2 - 6$
5) $17x^2 + 14$
6) $28x^2 + 13x$
7) $-8x^2 + 3x$
8) $8x^2 - 14x$
9) $-33x + 9$
10) $54x - 8$
11) $-18x - 64$
12) $7x^2 - 12x$
13) $-6x + 7$
14) $-9x + 15$
15) $32x + 13$
16) $-28x + 35$
17) $39x - 12$
18) $37x + 17$
19) $-2x + 32$
20) $-6x^2 - 5x$
21) $-48x^2 + 33x$
22) $-20x^2 + 73x$
23) $3x^2 + 44x$
24) $7x^2 - 49x - 28$
25) $3x^2 + 12x - 12$
26) $4x^2 - 21x$
27) $-43x^2 + 8x - 11$
28) $-5x^2 - 27x + 9$
29) $2x^2 + 18x + 28$
30) $19x^2 + 45x$
31) $12x^2 + 9x + 25$
32) $-2x^2 - 15x + 30$

Simplifying Polynomial Expressions

1) $5x^4 - 4x^2 - 11x$
2) $x^7 - 2x^4 - 4x^2$
3) $36x^5 + 7x^3$
4) $-37x^5 - 28x^2 + 14x$
5) $-9x^4 + 8x^2 - 5$
6) $-15x^5 + 9x^4 - 3x$
7) $-31x^3 - 5x^2 + 8x$
8) $-5x^3 - 5x^2 - 2x$
9) $-9x^6 + 8x^4 + 2x^2 - 2x$
10) $x^5 - 3x^3 + 3x$
11) $-6x^5 + 6x^2$
12) $-20x^5 + 17x^4 + 8x^3$
13) $9x^7 - 24x^6 + 4x^3$
14) $-13x^5 + 13x^3 + 24x$
15) $-9x^7 + 12x^6 - 12x^3$
16) $-17x^4 - 3x^3 + 19x$

Translate Phrases into an Algebraic Statement

1) $13x$
2) $y - 15$
3) $\frac{22}{x}$
4) $27 - y$
5) $y + 31$
6) 7^2
7) x^7
8) $5 + x$
9) $49 - y$
10) $\frac{8}{x}$
11) $\frac{x^2}{34}$
12) $x - 14 = 41$
13) $7b - a^2$
14) $51 - ab$

The Distributive Property

1) $20x + 8$
2) $20x + 10$
3) $30x - 30$
4) $-42x + 21$
5) $-4x - 32$
6) $24x + 24$
7) $35x - 40$
8) $12x + 3$

ACT Math Workbook

9) $40x - 15$	15) $63x - 36$	21) $18x - 8$	27) $56x - 35$
10) $-5x + 55$	16) $-36x - 60$	22) $-30x + 45$	28) $90x - 12$
11) $2x - 8$	17) $-15x + 45$	23) $-9x + 42$	29) $52x - 78$
12) $12x + 21$	18) $28x + 16$	24) $-50x + 20$	30) $60x - 52$
13) $12x + 32$	19) $36x - 72$	25) $84x - 60$	31) $-25x + 28$
14) $-40x + 10$	20) $-28x + 20$	26) $-16x - 72$	32) $-22x - 25$

Evaluating One Variables

1) 3	9) 17	17) 87	25) 76
2) −4	10) 4	18) −52	26) −54
3) 9	11) 12	19) 16	27) −85
4) −17	12) 6	20) 26	28) −130
5) 8	13) 30	21) 6	29) 15
6) 8	14) −1	22) −55	30) −1
7) 23	15) 27	23) −57	31) 8
8) −11	16) 18	24) 54	32) −12

Evaluating Two Variables

1) 17	6) −102	11) 3	16) 31
2) −2	7) 41	12) 71	17) 61
3) −13	8) 3	13) 16	18) −131
4) 27	9) 24	14) 15	19) 45
5) 31	10) 23	15) 48	20) 19

Combining like Terms

1) $9x + 8$	9) $2x - 9$	17) $4x + 15$	25) $-36x + 20$
2) $18x - 6$	10) $31x - 22$	18) $44x + 24$	26) $-46x + 14$
3) $-2x + 8$	11) $22x - 2$	19) $7x + 30$	27) $-89x + 30$
4) $-12x + 14$	12) $-32x + 16$	20) $15x - 12$	28) $8x - 20$
5) $10x - 12$	13) $14x + 30$	21) $19x + 30$	29) $-x + 15$
6) $22x - 22$	14) $-21x - 5$	22) $-32x + 18$	30) $52x - 17$
7) $-3x + 6$	15) $13x + 28$	23) $6x - 16$	31) $-41x + 26$
8) $-25x + 7$	16) $-8x - 27$	24) $-30x - 63$	32) $220x$

Chapter 6: Equations and Inequalities

Topics that you will practice in this chapter:

- ✓ One–Step Equations
- ✓ Multi–Step Equations
- ✓ Graphing Single–Variable Inequalities
- ✓ One–Step Inequalities
- ✓ Multi-Step Inequalities
- ✓ Systems of Equations
- ✓ Systems of Equations Word Problems

"Life is a math equation. In order to gain the most, you have to know how to convert negatives into positives." – Anonymous

One–Step Equations

✏ **Find the answer for each equation.**

1) $3x = 90$, $x =$ ___

2) $5x = 35$, $x =$ ___

3) $9x = 36$, $x =$ ___

4) $25x = 150$, $x =$ ___

5) $x + 18 = 23$, $x =$ ___

6) $x - 3 = 8$, $x =$ ___

7) $x - 7 = 4$, $x =$ ___

8) $x + 22 = 30$, $x =$ ___

9) $x - 11 = 6$, $x =$ ___

10) $24 = 28 + x$, $x =$ ___

11) $x - 5 = 7$, $x =$ ___

12) $9 - x = -7$, $x =$ ___

13) $43 = -8 + x$, $x =$ ___

14) $x - 23 = -38$, $x =$ ___

15) $x + 45 = -27$, $x =$ ___

16) $42 = 56 - x$, $x =$ ___

17) $-18 + x = -32$, $x =$ ___

18) $x - 13 = 7$, $x =$ ___

19) $35 = x - 10$, $x =$ ___

20) $x - 8 = -21$, $x =$ ___

21) $x - 54 = -20$, $x =$ ___

22) $x - 42 = -47$, $x =$ ___

23) $x - 8 = 29$, $x =$ ___

24) $-93 = x - 51$, $x =$ ___

25) $x + 15 = 37$, $x =$ ___

26) $108 = 12x$, $x =$ ___

27) $x - 33 = 27$, $x =$ ___

28) $x - 12 = 23$, $x =$ ___

29) $72 - x = 18$, $x =$ ___

30) $x + 34 = 58$, $x =$ ___

31) $21 - x = -9$, $x =$ ___

32) $x - 59 = -80$, $x =$ ___

Multi-Step Equations

✏️ **Find the answer for each equation.**

1) $3x + 1 = 7$

2) $-x + 10 = 9$

3) $5x - 13 = 7$

4) $-(4 - x) = 5$

5) $3x - 8 = 16$

6) $15x - 13 = 17$

7) $3x - 28 = 2$

8) $9x + 21 = 39$

9) $14x + 17 = 45$

10) $-14(8 + x) = 70$

11) $8(10 + x) = 32$

12) $16 = -(x - 8)$

13) $5(7 - 3x) = 50$

14) $-19 = -(3x + 7)$

15) $30(3 + x) = 60$

16) $9(x - 12) = 54$

17) $-24 = 3x + 5x$

18) $5x + 28 = -2x - 7$

19) $9(5 + 4x) = -99$

20) $18 - x = -12 - 6x$

21) $4 - 4x = 28 - 2x$

22) $15 + 12x = -15 + 8x$

23) $54 = (-3x) - 8 + 8$

24) $12 = 7x - 18 + 5x$

25) $-18 = -9x - 42 + 5x$

26) $11x - 6 = -33 + 8x$

27) $8x - 42 = 3x + 3$

28) $-15 - 8x = 4(5 - x)$

29) $x - 9 = -5(9 - 2x)$

30) $14x - 65 = -x - 110$

31) $3x - 129 = -3(11 + 7x)$

32) $-7x - 20 = 2x + 43$

Graphing Single–Variable Inequalities

✏ **Draw a graph for each inequality.**

1) $x \leq 7$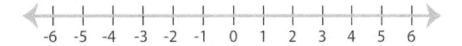

2) $x \leq -1.5$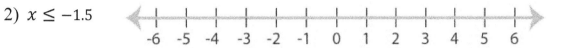

3) $x < -4$

4) $x > 2.5$

5) $x > 1.3$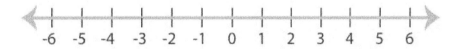

6) $x < 4$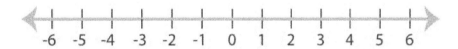

7) $x < 2.4$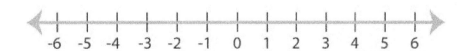

8) $x > -\frac{18}{10}$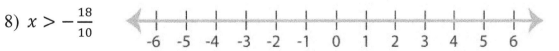

One–Step Inequalities

✏️ Find the answer for each inequality and graph it.

1) $x + 3 > -5$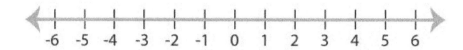

2) $x - 4 < 1$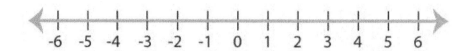

3) $7x < 42$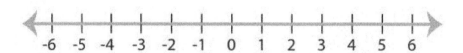

4) $13 + x > 12$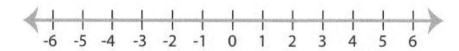

5) $x + 20 < 13$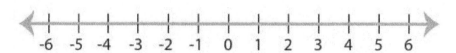

6) $14x \le 42$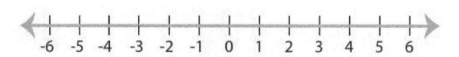

7) $11x \le -44$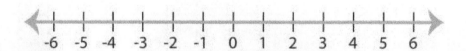

8) $x + 26 > 35$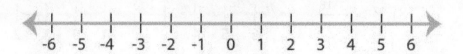

Multi-Step Inequalities

✎ **Calculate each inequality.**

1) $x - 8 \leq 12$

2) $9 - 3x \leq 18$

3) $4x - 7 \leq 9$

4) $8x - 9 \geq 15$

5) $x - 19 \geq 24$

6) $5x - 15 \leq 40$

7) $7x - 4 \leq 24$

8) $-18 + 8x \leq 22$

9) $9(x - 8) \leq 27$

10) $4x - 8 \leq 16$

11) $11x - 42 < 22$

12) $10x - 18 < 52$

13) $17 - 9x \geq -46$

14) $32 + 2x < 68$

15) $8 + 8x \geq 80$

16) $11 + 6x < 65$

17) $9x - 13 < 23$

18) $8(12 - 4x) \geq -68$

19) $-(2 + 5x) < 42$

20) $14 - 9x \geq -31$

21) $-5(x - 3) > 65$

22) $\dfrac{2x + 8}{3} \leq 12$

23) $\dfrac{8x + 16}{4} \leq 24$

24) $\dfrac{2x - 22}{9} > 8$

25) $7 + \dfrac{x}{4} < 21$

26) $\dfrac{32x}{16} - 4 < 6$

27) $\dfrac{12x + 36}{22} > 3$

28) $42 + \dfrac{x}{3} < 15$

ACT Math Workbook

Systems of Equations

✎ **Calculate each system of equations.**

1) $-6x + 7y = 8$ $x = $ ___
   $x + 4y = 9$ $y = $ ___

2) $-4x + 12y = 12$ $x = $ ___
   $14x - 16y = 10$ $y = $ ___

3) $y = -9$ $x = $ ___
   $2x - 5y = 12$ $y = $ ___

4) $4y = -4x + 20$ $x = $ ___
   $8x - 2y = -12$ $y = $ ___

5) $10x - 9y = -13$ $x = $ ___
   $-5x + 3y = 11$ $y = $ ___

6) $-6x - 8y = 10$ $x = $ ___
   $4x - 8y = 20$ $y = $ ___

7) $5x - 14y = -23$ $x = $ ___
   $-6x + 7y = 8$ $y = $ ___

8) $-4x + 3y = 3$ $x = $ ___
   $-x + 2y = 5$ $y = $ ___

9) $-4x + 5y = 15$ $x = $ ___
   $-3x + 4y = -10$ $y = $ ___

10) $-6x - 6y = -21$ $x = $ ___
    $-6x + 6y = -66$ $y = $ ___

11) $12x - 21y = 6$ $x = $ ___
    $-6x - 3y = -12$ $y = $ ___

12) $-4x - 4y = -14$ $x = $ ___
    $4x - 4y = 44$ $y = $ ___

13) $4x + 5y = 3$ $x = $ ___
    $3x - y = 6$ $y = $ ___

14) $3x - 2y = 2$ $x = $ ___
    $10x - 10y = 20$ $y = $ ___

15) $5x + 8y = 14$ $x = $ ___
    $-3x - 2y = -3$ $y = $ ___

16) $8x + 5y = 4$ $x = $ ___
    $-3x - 4y = 15$ $y = $ ___

WWW.MathNotion.Com

Systems of Equations Word Problems

✎ **Find the answer for each word problem.**

1) Tickets to a movie cost $6 for adults and $4 for students. A group of friends purchased 9 tickets for $50.00. How many adults ticket did they buy? ____

2) At a store, Eva bought two shirts and five hats for $77.00. Nicole bought three same shirts and four same hats for $84.00. What is the price of each shirt? _____

3) A farmhouse shelters 10 animals, some are pigs, and some are ducks. Altogether there are 36 legs. How many pigs are there? ____

4) A class of 85 students went on a field trip. They took 24 vehicles, some cars and some buses. If each car holds 3 students and each bus hold 16 students, how many buses did they take? _____

5) A theater is selling tickets for a performance. Mr. Smith purchased 8 senior tickets and 10 child tickets for $248 for his friends and family. Mr. Jackson purchased 4 senior tickets and 6 child tickets for $132. What is the price of a senior ticket? $____

6) The difference of two numbers is 15. Their sum is 33. What is the bigger number? $____

7) The sum of the digits of a certain two-digit number is 7. Reversing its digits increase the number by 9. What is the number? _____

8) The difference of two numbers is 11. Their sum is 25. What are the numbers? _____

9) The length of a rectangle is 5 meters greater than 2 times the width. The perimeter of rectangle is 28 meters. What is the length of the rectangle? _____

10) Jim has 23 nickels and dimes totaling $2.40. How many nickels does he have? ____

Answers of Worksheets – Chapter 6

One–Step Equations

1) 30	9) 17	17) −14	25) 22
2) 7	10) −4	18) 20	26) 9
3) 4	11) 12	19) 45	27) 60
4) 6	12) 16	20) −13	28) 35
5) 5	13) 51	21) 34	29) 54
6) 11	14) −15	22) −5	30) 24
7) 11	15) −72	23) 37	31) 30
8) 8	16) 14	24) −42	32) −21

Multi–Step Equations

1) 2	9) 2	17) −3	25) −6
2) 1	10) −13	18) −5	26) −9
3) 4	11) −6	19) −4	27) 9
4) 9	12) −8	20) −6	28) −8.75
5) 8	13) −1	21) −12	29) 4
6) 2	14) 4	22) −7.5	30) −3
7) 10	15) −1	23) −18	31) 4
8) 2	16) 18	24) 2.5	32) −7

Graphing Single–Variable Inequalities

1) [number line with point at 7]

2) [number line with point at −1.5]

3) [number line with open circle at −4]

4) [number line with open circle at 2.5]

5)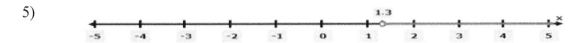

6)

7)

8)

One–Step Inequalities

1)

2)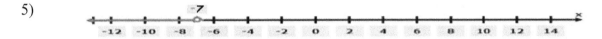

3)

4)

5)

6)

7)

8)

Multi-Step Inequalities

1) $x \leq 20$	5) $x \geq 43$	9) $x \leq 11$	13) $x \leq 7$
2) $x \geq -3$	6) $x \leq 11$	10) $x \leq 6$	14) $x < 18$
3) $x \leq 4$	7) $x \leq 4$	11) $x < 64/11$	15) $x \geq 9$
4) $x \geq 3$	8) $x \leq 5$	12) $x < 7$	16) $x < 9$

ACT Math Workbook

17) $x < 4$
18) $x \leq 41/8$
19) $x > -44/5$
20) $x \leq 5$
21) $x < -10$
22) $x \leq 14$
23) $x \leq 10$
24) $x > 47$
25) $x < 56$
26) $x < 9/4$
27) $x > 2.5$
28) $x < -81$

Systems of Equations

1) $x = 1, y = 2$
2) $x = 3, y = 2$
3) $x = -\frac{33}{2}$
4) $x = -\frac{1}{5}, y = \frac{26}{5}$
5) $x = -4, y = -3$
6) $x = 1, y = -2$
7) $x = 1, y = 2$
8) $x = \frac{9}{5}, y = \frac{17}{5}$
9) $x = -110, y = -85$
10) $x = -\frac{15}{4}, y = \frac{29}{4}$
11) $x = \frac{5}{3}, y = \frac{2}{3}$
12) $x = -\frac{15}{4}, y = \frac{29}{4}$
13) $x = \frac{33}{19}, y = -\frac{15}{19}$
14) $x = -2, y = -4$
15) $x = -\frac{2}{7}, y = \frac{27}{14}$
16) $x = \frac{91}{17}, y = -\frac{132}{17}$

Systems of Equations Word Problems

1) 7
2) $16
3) 8
4) 1
5) $21
6) 24
7) 43
8) 18, 7
9) 11 meters
10) 18

Chapter 7:
Linear Functions

Topics that you will practice in this chapter:

- ✓ Finding Slope
- ✓ Graphing Lines Using Line Equation
- ✓ Writing Linear Equations
- ✓ Graphing Linear Inequalities
- ✓ Finding Midpoint
- ✓ Finding Distance of Two Points

"Nature is written in mathematical language." – Galileo Galilei

Finding Slope

✎ **Find the slope of each line.**

1) $y = 2x + 5$

2) $y = -x + 17$

3) $y = 4x + 16$

4) $y = -3x + 15$

5) $y = 27 + 7x$

6) $y = 11 - 4x$

7) $y = 7x + 14$

8) $y = -8x + 18$

9) $y = -9x + 15$

10) $y = 8x - 13$

11) $y = \frac{1}{5}x + 9$

12) $y = -\frac{3}{7}x + 19$

13) $-3x + 6y = 17$

14) $4x + 4y = 16$

15) $8y - 3x = 32$

16) $11y - 3x = 42$

✎ **Find the slope of the line through each pair of points.**

17) $(1, 8), (5, 16)$

18) $(-2, 14), (2, 18)$

19) $(7, -1), (3, 9)$

20) $(-4, -4), (2, 14)$

21) $(16, -1), (4, 11)$

22) $(-21, 5), (-10, 38)$

23) $(8, 11), (12, 19)$

24) $(22, -22), (10, 14)$

25) $(21, -15), (19, -13)$

26) $(11, 10), (7, -2)$

27) $(5, 4), (9, 16)$

28) $(34, -87), (22, 45)$

Graphing Lines Using Line Equation

✏️ **Sketch the graph of each line.**

1) $y = x - 5$ 2) $y = -3x + 4$ 3) $x - 2y = 0$

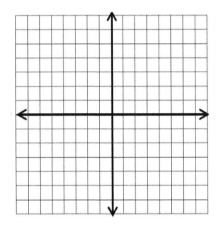 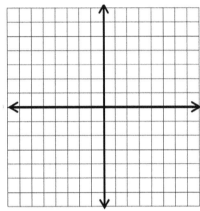 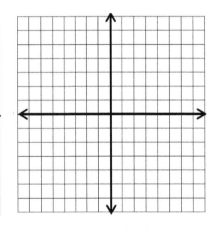

4) $x + y = -4$ 5) $4x + 3y = -2$ 6) $y - 3x + 6 = 0$

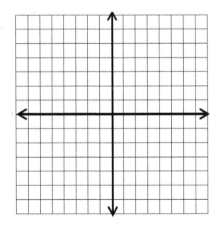 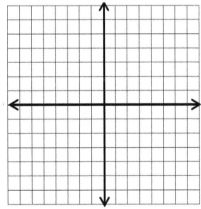 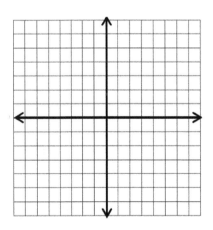

Writing Linear Equations

✎ **Write the equation of the line through the given points.**

1) Through: $(6, -10), (10, 14)$

2) Through: $(10, 4), (4, 22)$

3) Through: $(-6, 4), (2, 12)$

4) Through: $(15, 11), (3, -1)$

5) Through: $(-5, 33), (9, 5)$

6) Through: $(20, 5), (17, 2)$

7) Through: $(24, -4), (16, 4)$

8) Through: $(-18, 57), (33, -45)$

9) Through: $(10, 12), (8, 18)$

10) Through: $(25, 41), (33, -7)$

11) Through: $(-6, 9), (-8, -7)$

12) Through: $(8, 8), (4, -8)$

13) Through: $(6, -10), (10, 6)$

14) Through: $(10, -24), (-8, 12)$

15) Through: $(10, 10), (-2, -4)$

16) Through: $(-7, 35), (11, -31)$

✎ **Find the answer for each problem.**

17) What is the equation of a line with slope 3 and intercept 11? _____

18) What is the equation of a line with slope 5 and intercept 15? _____

19) What is the equation of a line with slope 7 and passes through point $(3, 2)$? _____

20) What is the equation of a line with slope -3 and passes through point $(-2, 5)$? _____

21) The slope of a line is -6 and it passes through point $(-2, 1)$. What is the equation of the line? _____

22) The slope of a line is 5 and it passes through point $(-4, 2)$. What is the equation of the line? _____

Graphing Linear Inequalities

✎ **Sketch the graph of each linear inequality.**

1) $y > 3x - 5$
2) $y < 2x + 1$
3) $y \leq -4x - 5$

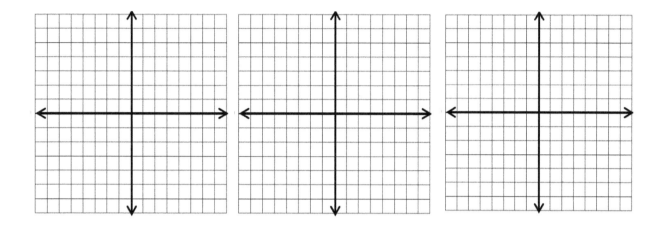

4) $2y \geq 12 + 4x$
5) $-5y < x - 15$
6) $3y \geq -9x + 6$

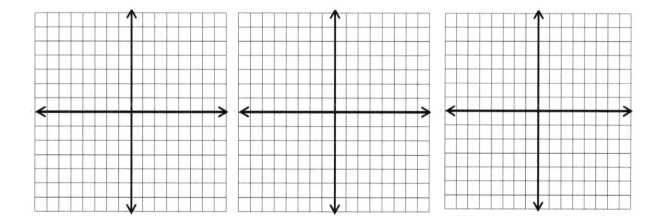

Finding Midpoint

✏️ **Find the midpoint of the line segment with the given endpoints.**

1) $(-4, -6), (2, 4)$

2) $(13, 5), (-1, 5)$

3) $(11, -4), (3, 14)$

4) $(-15, -6), (3, 9)$

5) $(7, -8), (13, -12)$

6) $(-14, -8), (8, -12)$

7) $(9, 2), (-9, 22)$

8) $(-8, 10), (-8, 4)$

9) $(-7, 7), (23, -15)$

10) $(3, 17), (19, -5)$

11) $(-4, 13), (7, 9)$

12) $(11, 8), (-3, -6)$

13) $(-4, 12), (0, 6)$

14) $(34, 12), (18, -28)$

15) $(15, 6), (-1, 0)$

16) $(-11, -13), (-13, 19)$

17) $(12, 4), (8, 16)$

18) $(-2, -7), (18, -21)$

19) $(18, 13), (-6, 5)$

20) $(10, -4), (0, 18)$

21) $(4, -4), (8, -20)$

22) $(25, 5), (-11, -17)$

23) $(8, 12), (16, -2)$

24) $(14, -20), (8, 14)$

✏️ **Find the answer for each problem.**

25) One endpoint of a line segment is $(6, 8)$ and the midpoint of the line segment is $(1, 6)$. What is the other endpoint? _____

26) One endpoint of a line segment is $(-7, 5)$ and the midpoint of the line segment is $(1, 3)$. What is the other endpoint? _____

27) One endpoint of a line segment is $(-6, -10)$ and the midpoint of the line segment is $(2, 9)$. What is the other endpoint? _____

Finding Distance of Two Points

◆ **Find the distance between each pair of points.**

1) $(5, 9), (-11, -3)$

2) $(-6, 2), (-2, 6)$

3) $(-8, -1), (-3, 8)$

4) $(-8, -2), (2, 22)$

5) $(6, -4), (-12, -28)$

6) $(-6, 0), (-2, 3)$

7) $(8, 12), (8, 6)$

8) $(12, -10), (12, -2)$

9) $(15, 27), (-33, -9)$

10) $(10, -2), (6, -14)$

11) $(1, 0), (6, 12)$

12) $(8, 4), (3, -8)$

13) $(3, 2), (-5, -11)$

14) $(-10, 12), (6, 42)$

15) $(0, 16), (-8, 10)$

16) $(5, 0), (30, 60)$

17) $(3, 5), (-5, -10)$

18) $(-4, 6), (4, 3)$

19) $(7, 2), (-8, -18)$

20) $(-10, 8), (14, 18)$

◆ **Find the answer for each problem.**

21) Triangle ABC is a right triangle on the coordinate system and its vertices are $(-5, 7), (-5, 1)$, and $(1, 1)$. What is the area of triangle ABC? _____

22) Three vertices of a triangle on a coordinate system are $(1, 1), (7, 1)$, and $(1, 9)$. What is the perimeter of the triangle? _____

23) Four vertices of a rectangle on a coordinate system are $(-2, 4), (-2, 7), (4, 4)$, and $(4, 7)$. What is its perimeter? _____

Answers of Worksheets – Chapter 7

Finding Slope

1) 2
2) -1
3) 4
4) -3
5) 7
6) -4
7) 7
8) -8
9) -9
10) 8
11) $\frac{1}{5}$
12) $-\frac{3}{7}$
13) $\frac{1}{2}$
14) -1
15) $\frac{3}{8}$
16) $\frac{3}{11}$
17) 2
18) 1
19) $-\frac{5}{2}$
20) 3
21) -1
22) 3
23) 2
24) -3
25) -1
26) 3
27) 3
28) -11

Graphing Lines Using Line Equation

1) $y = x - 5$

2) $y = -3x + 4$

3) $x - 2y = 0$

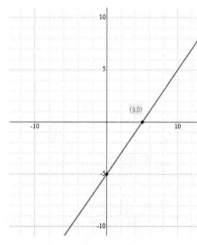
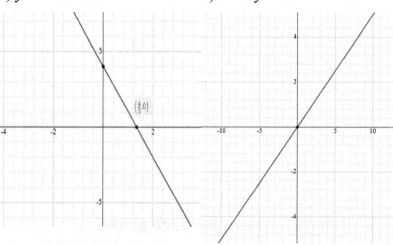

4) $x + y = -4$

5) $4x + 3y = -2$

6) $y - 3x + 6 = 0$

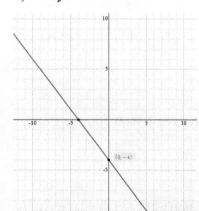
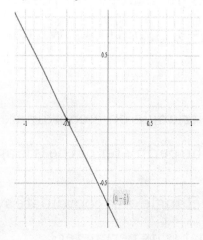
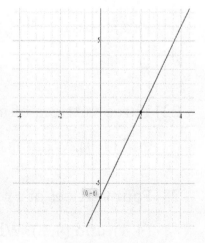

ACT Math Workbook

Writing Linear Equations

1) $y = 6x - 46$
2) $y = -3x + 34$
3) $y = x + 10$
4) $y = x - 4$
5) $y = -2x + 23$
6) $y = x - 15$
7) $y = -x + 20$
8) $y = -2x + 21$
9) $y = -3x + 42$
10) $y = -6x + 191$
11) $y = 8x + 57$
12) $y = 4x - 24$
13) $y = 4x - 34$
14) $y = -2x - 4$
15) $y = \frac{7}{6}x - \frac{5}{3}$
16) $y = -\frac{11}{3}x + \frac{28}{3}$
17) $y = 3x + 11$
18) $y = 5x + 15$
19) $y = 7x - 19$
20) $y = -3x - 1$
21) $y = -6x - 11$
22) $y = 5x + 22$

Graphing Linear Inequalities

1) $y > 3x - 5$

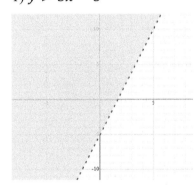

2) $y < 2x + 1$

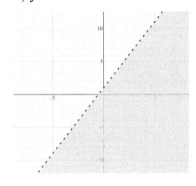

3) $y \leq -4x - 5$

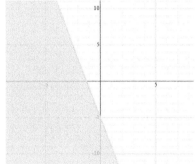

4) $2y \geq 12 + 4x$

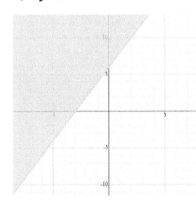

5) $-5y < x - 15$

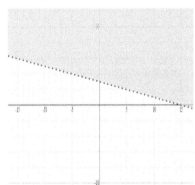

6) $3y \geq -9x + 6$

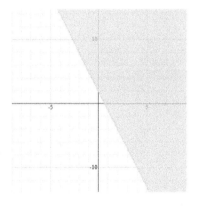

Finding Midpoint

1) $(-1, -1)$
2) $(6, 5)$
3) $(7, 5)$
4) $(-6, 1.5)$
5) $(10, -10)$
6) $(-3, -10)$
7) $(0, 12)$
8) $(-8, 7)$
9) $(8, -4)$

ACT Math Workbook

10) (11, 6)	16) (−12, 3)	22) (7, −6)
11) (1.5, 11)	17) (10, 10)	23) (12, 5)
12) (4, 1)	18) (8, −14)	24) (11, −3)
13) (−2, 9)	19) (6, 9)	25) (−4, 4)
14) (26, −8)	20) (5, 7)	26) (9, 1)
15) (7, 3)	21) (6, −12)	27) (10, 28)

Finding Distance of Two Points

1) 20	9) 60	17) 17
2) $4\sqrt{2}$	10) $4\sqrt{10}$	18) $\sqrt{73}$
3) $\sqrt{106}$	11) 13	19) 25
4) 26	12) 13	20) 26
5) 30	13) $\sqrt{233}$	21) 18 square units
6) 10	14) 34	22) 24 units
7) 6	15) 10	23) 18 units
8) 8	16) 65	

Chapter 8:

Polynomials

Topics that you will practice in this chapter:

- ✓ Writing Polynomials in Standard Form
- ✓ Simplifying Polynomials
- ✓ Adding and Subtracting Polynomials
- ✓ Multiplying Monomials
- ✓ Multiplying and Dividing Monomials
- ✓ Multiplying a Polynomial and a Monomial
- ✓ Multiplying Binomials
- ✓ Factoring Trinomials
- ✓ Operations with Polynomials

Mathematics is the supreme judge; from its decisions there is no appeal. – Tobias Dantzig

Writing Polynomials in Standard Form

✍ Write each polynomial in standard form.

1) $9x - 7x =$

2) $-6 + 15x - 15x =$

3) $3x^2 - 11x^3 =$

4) $18 + 19x^3 - 14 =$

5) $3x^2 + 9x - 4x^5 =$

6) $-7x^3 + 12x^7 =$

7) $9x + 6x^2 - 2x^6 =$

8) $-5x^3 + x - 9x^4 =$

9) $8x^2 + 34 - 21x =$

10) $8 - 7x + 11x^4 =$

11) $25x^3 + 45x - 13x^4 =$

12) $17 + 9x^2 - 2x^3 =$

13) $18x^2 - 8x + 8x^3 =$

14) $9x^4 - 4x^2 - 10x^5 =$

15) $-41 + 7x^2 - 8x^4 =$

16) $8x^2 - 7x^5 + 3x^3 - 12 =$

17) $4x^2 - 9x^5 + 12 - 8x^4 =$

18) $-2x^5 + 6x - 9x^2 - 7x =$

19) $14x^5 + 7x^4 - 8x^5 - 8x^2 =$

20) $2x^3 - 15x^4 + 9x^3 + 3x^8 =$

21) $7x^4 - 16x^5 - 9x^2 + 10x^4 =$

22) $5x^2 + 6x^5 + 37x^3 - 9x^5 =$

23) $3x(2x + 5 - 6x^2) =$

24) $12x(x^6 + 2x^3) =$

25) $6x(x^2 + 8x + 4) =$

26) $8x(3 - 2x + 4x^3) =$

27) $7x(2x^3 - 2x^2 + 2) =$

28) $5x(5x^5 + 4x^4 - 1) =$

29) $x(4x^3 + 52x^4 + 2x) =$

30) $6x(3x - 4x^4 + 7x^2) =$

Simplifying Polynomials

✎ **Simplify each expression.**

1) $3(x - 12) =$

2) $5x(2x - 4) =$

3) $7x(5x - 1) =$

4) $6x(3x + 2) =$

5) $5x(2x - 7) =$

6) $9x(x + 8) =$

7) $(3x - 8)(x - 3) =$

8) $(x - 9)(3x + 4) =$

9) $(x - 8)(x - 5) =$

10) $(3x + 4)(3x - 4) =$

11) $(5x - 8)(5x - 2) =$

12) $7x^2 + 7x^2 - 6x^4 =$

13) $5x - 2x^2 + 7x^3 + 10 =$

14) $8x + 2x^2 - 5x^3 =$

15) $15x + 4x^5 - 8x^2 =$

16) $-4x^2 + 7x^5 + 11x^4 =$

17) $-14x^2 + 8x^3 - 2x^4 + 5x =$

18) $14 - 5x^2 + 6x^2 - 10x^3 + 17 =$

19) $x^2 - 9x + 2x^3 + 15x - 10x =$

20) $14 - 8x^2 + 4x^2 - 9x^3 + 1 =$

21) $-4x^5 + 2x^4 - 18x^2 + 2x^5 =$

22) $(3x^3 - 5) + (3x^3 - 2x^3) =$

23) $4(3x^5 - 3x^3 - 6x^5) =$

24) $-4(x^5 + 8) - 4(12 - x^5) =$

25) $7x^2 - 9x^3 - 2x + 14 - 5x^2 =$

26) $10 - 5x^2 + 3x^2 - 4x^3 + 4 =$

27) $(8x^2 - 2x) - (5x - 5 - 4x^2) =$

28) $4x^4 - 8x^3 - x(3x^2 + 5x) =$

29) $4x + 8x^2 - 10 - 2(x^2 - 1) =$

30) $5 - 3x^2 + (6x^4 - 2x^2 + 8x^4) =$

31) $-(x^5 + 8) - 7(4 + x^5) =$

32) $(4x^3 - x) - (x - 6x^3) =$

Adding and Subtracting Polynomials

✎ **Add or subtract expressions.**

1) $(-x^3 - 3) + (4x^3 + 2) =$

2) $(3x^2 + 4) - (6 - x^2) =$

3) $(x^3 + 4x^2) - (5x^3 + 15) =$

4) $(3x^3 - 2x^2) + (2x^2 - x) =$

5) $(10x^3 + 14x) - (14x^3 + 7) =$

6) $(5x^2 - 7) + (3x^2 + 7) =$

7) $(9x^3 + 4) - (10 - 5x^3) =$

8) $(x^2 + 2x^3) - (2x^3 + 5) =$

9) $(8x^2 - x) + (5x - 4x^2) =$

10) $(17x + 10) - (2x + 10) =$

11) $(12x^4 - 4x) - (x - 3x^4) =$

12) $(3x - x^4) - (7x^4 + 8x) =$

13) $(7x^3 - 6x^5) - (4x^5 - 2x) =$

14) $(x^3 - 7) + (4x^3 + 8x^5) =$

15) $(6x^2 + 5x^4) - (x^4 - 9x^2) =$

16) $(-4x^2 - 4x) + (7x - 8x^2) =$

17) $(x - 6x^4) - (15x^4 + 2x) =$

18) $(4x - 3x^4) - (2x^4 - 3x^3) =$

19) $(7x^3 - 7) + (6x^3 - 6x^2) =$

20) $(9x^5 + 7x^4) - (x^4 - 5x^5) =$

21) $(-4x^2 + 11x^4 + 2x^3) + (20x^3 + 4x^4 + 12x^2) =$

22) $(5x^2 - 5x^4 - 5x) - (-4x^2 - 5x^4 + 5x) =$

23) $(12x + 36x^3 - 10x^4) + (20x^3 + 10x^4 - 7x) =$

24) $(2x^5 - 4x^3 - 5x) - (2x^2 + 7x^3 - 2x) =$

25) $(14x^3 - 4x^5 - x) - (-4x^3 - 12x^5 + 9x) =$

26) $(-5x^2 + 12x^4 + x^3) + (10x^3 + 17x^4 + 7x^2) =$

Multiplying Monomials

✎ **Simplify each expression.**

1) $7u^5 \times (-u^2) =$

2) $(-9p^8) \times (-4p^2) =$

3) $5xy^3z^3 \times 4z^2 =$

4) $8u^6t \times 2ut^2 =$

5) $(-2a^2) \times (-5a^3b^3) =$

6) $-4a^2b^2 \times 5a^4b =$

7) $10xy^4 \times 2x^2y^2 =$

8) $4p^2q^4 \times (-2pq^2) =$

9) $8s^5t^4 \times 3st^4 =$

10) $(-7x^5y^3) \times 7x^4y =$

11) $xy^7z \times 15z^3 =$

12) $15xy \times 2x^3y =$

13) $14pq^4 \times (-3p^3q) =$

14) $25s^4t^2 \times st^6 =$

15) $12p^5 \times (-2p^3) =$

16) $(-12p^2q^4r) \times 3pq^5r^3 =$

17) $(-7a^4) \times (-4a^5b) =$

18) $4u^7v^2 \times (-9u^4v^6) =$

19) $9u^5 \times (-3u) =$

20) $-3xy^9 \times 8x^5y =$

21) $12y^5z^3 \times (-2y^2z) =$

22) $9a^3bc^5 \times 4abc^3 =$

23) $(-9p^5q^2) \times (-3p^2q^4) =$

24) $4u^8v^3 \times (-4u^8v^5) =$

25) $15y^3z^4 \times (-y^5z) =$

26) $(-12pq^4r^3) \times 5p^4q^2r =$

27) $3ab^5c^2 \times 3a^2bc^4 =$

28) $7x^5yz^3 \times 9x^5y^7z^4 =$

Multiplying and Dividing Monomials

✎ **Simplify each expression.**

1) $(7x^2)(x^3) =$

2) $(4x^3)(5x^2) =$

3) $(3x^4)(2x^2) =$

4) $(5x^8)(8x^3) =$

5) $(12x^6)(2x^3) =$

6) $(2yx^5)(16x^2) =$

7) $(9x^5y)(2x^2y^3) =$

8) $(-2x^2y^5)(5x^3y^2) =$

9) $(-4x^2y^2)(-8x^4y^3) =$

10) $(2x^4y)(-5x^5y^3) =$

11) $(9x^4y^4)(2x^3y^3) =$

12) $(2x^4y^6)(3x^4y^3) =$

13) $(8x^3y^8)(7x^5y^{10}) =$

14) $(14x^6y^5)(3x^5y^5) =$

15) $(8x^2y^8)(5x^{10}y^{10}) =$

16) $(-3x^2y^5)(4x^6y^3) =$

17) $\dfrac{9x^4y^5}{xy^3} =$

18) $\dfrac{18x^8y^3}{18x^7y} =$

19) $\dfrac{54x^4y^4}{6xy} =$

20) $\dfrac{63x^4y^5}{7x^3y^4} =$

21) $\dfrac{32x^7y^6}{8x^2y^3} =$

22) $\dfrac{63x^9y^4}{3x^4y^3} =$

23) $\dfrac{96x^{16}y^{12}}{12x^7y^9} =$

24) $\dfrac{60x^{10}y^6}{12x^{11}y^3} =$

25) $\dfrac{90x^8y^{12}}{18x^7y^{12}} =$

26) $\dfrac{45x^{23}y^{10}}{9x^9y^6} =$

27) $\dfrac{-96x^8y^8}{24x^6y^8} =$

Multiplying a Polynomial and a Monomial

✎ **Find each product.**

1) $2x(x + 4) =$

2) $3(8 - x) =$

3) $5x(3x + 4) =$

4) $x(-2x + 5) =$

5) $7x(3x - 3) =$

6) $3(2x - 5y) =$

7) $6x(7x - 3) =$

8) $x(12x + 5y) =$

9) $5x(x + 6y) =$

10) $11x(4x + 5y) =$

11) $8x(4x + 2) =$

12) $12x(x - 15y) =$

13) $9x(5x - 3y) =$

14) $8x(5x - 2y + 5) =$

15) $9x(2x^2 + 7y^2) =$

16) $8x(9x + 6y) =$

17) $2(3x^5 - 2y^5) =$

18) $4x(-x^2y + 2y) =$

19) $-3(2x^3 - 3xy + 9) =$

20) $2(x^2 - 2xy - 4) =$

21) $7x(4x^3 - xy + 2x) =$

22) $-9x(-2x^3 - 2x + 7xy) =$

23) $6(x^2 + 3xy - 8y^2) =$

24) $5x(7x^3 - x + 8) =$

25) $7(x^{24} - 4x - 6) =$

26) $x^2(-3x^3 + 4x + 7) =$

27) $x^2(2x^3 + 10 - 5x) =$

28) $4x^4(3x^3 - 2x + 8) =$

29) $5x^2(x^4 - 5xy + 2y^3) =$

30) $4x^2(7x^4 - 2x + 11) =$

31) $7x^3(3x^3 + 5x - 7) =$

32) $4x(x^2 - 8xy + 7y^3) =$

Multiplying Binomials

✏ **Find each product.**

1) $(x+5)(x+1) =$

2) $(x-3)(x+7) =$

3) $(x-1)(x-9) =$

4) $(x+3)(x+8) =$

5) $(x-4)(x-11) =$

6) $(x+5)(x+6) =$

7) $(x-8)(x+7) =$

8) $(x-3)(x-2) =$

9) $(x+8)(x+11) =$

10) $(x-3)(x+5) =$

11) $(x+8)(x+8) =$

12) $(x+2)(x+7) =$

13) $(x-9)(x+4) =$

14) $(x-10)(x+10) =$

15) $(x+24)(x+2) =$

16) $(x+9)(x+13) =$

17) $(x-7)(x+7) =$

18) $(x-5)(x+2) =$

19) $(3x+4)(x+5) =$

20) $(x-8)(5x+2) =$

21) $(x-9)(4x+9) =$

22) $(2x-7)(3x-2) =$

23) $(x-4)(x+11) =$

24) $(5x-6)(2x+4) =$

25) $(4x-9)(x+7) =$

26) $(8x-5)(2x+2) =$

27) $(3x+9)(7x+4) =$

28) $(6x-8)(4x+4) =$

29) $(4x+5)(5x-8) =$

30) $(8x-1)(8x+4) =$

31) $(9x+4)(3x-6) =$

32) $(4x^2+12)(4x^2-12) =$

Factoring Trinomials

✎ **Factor each trinomial.**

1) $x^2 + 12x + 35 =$

2) $x^2 - 8x + 12 =$

3) $x^2 + 11x + 10 =$

4) $x^2 - 12x + 27 =$

5) $x^2 - 16x + 15 =$

6) $x^2 - 13x + 40 =$

7) $x^2 + 15x + 44 =$

8) $x^2 + x - 72 =$

9) $x^2 - 81 =$

10) $x^2 - 17x + 70 =$

11) $x^2 + 8x - 48 =$

12) $x^2 + 5x - 104 =$

13) $x^2 - 7x - 18 =$

14) $x^2 + 22x + 121 =$

15) $3x^2 - 3x - 36 =$

16) $2x^2 - 35x + 75 =$

17) $14x^2 + 11x - 15 =$

18) $8x^2 - 12x - 20 =$

19) $15x^2 + 16x + 4 =$

20) $24x^2 + 2x - 1 =$

✎ **Calculate each problem.**

21) The area of a rectangle is $x^2 - 3x - 40$. If the width of rectangle is $x - 8$, what is its length? _____

22) The area of a parallelogram is $12x^2 + 7x - 10$ and its height is $4x + 5$. What is the base of the parallelogram? _____

23) The area of a rectangle is $10x^2 - 43x + 28$. If the width of the rectangle is $5x - 4$, what is its length? _____

Operations with Polynomials

✏️ **Find each product.**

1) $2(4x + 1) = $ _____

2) $5(2x + 7) = $ _____

3) $4(6x - 5) = $ _____

4) $-4(7x - 8) = $ _____

5) $3x^2(8x + 4) = $ _____

6) $6x^2(2x - 9) = $ _____

7) $5x^3(-x + 4) = $ _____

8) $-5x^4(4x - 9) = $ _____

9) $6(x^2 + 7x - 3) = $ _____

10) $4(3x^2 - 2x + 6) = $ _____

11) $9(3x^2 + 8x + 2) = $ _____

12) $7x(x^2 + 5x + 3) = $ _____

13) $(7x + 2)(2x - 5) = $ _____

14) $(8x + 5)(3x - 8) = $ _____

15) $(4x + 2)(6x - 1) = $ _____

16) $(5x - 4)(5x + 9) = $ _____

✏️ **Calculate each problem.**

17) The measures of two sides of a triangle are $(2x + 8y)$ and $(5x - 3y)$. If the perimeter of the triangle is $(11x + 6y)$, what is the measure of the third side? _____

18) The height of a triangle is $(8x + 2)$ and its base is $(2x - 6)$. What is the area of the triangle? _____

19) One side of a square is $(4x + 3)$. What is the area of the square? _____

20) The length of a rectangle is $(7x - 9y)$ and its width is $(13x + 9y)$. What is the perimeter of the rectangle? _____

21) The side of a cube measures $(x + 2)$. What is the volume of the cube? _____

22) If the perimeter of a rectangle is $(24x + 10y)$ and its width is $(4x + 3y)$, what is the length of the rectangle? _____

Answers of Worksheets – Chapter 8

Writing Polynomials in Standard Form

1) $2x$
2) -6
3) $-11x^3 + 3x^2$
4) $19x^4 + 4$
5) $-4x^5 + 3x^2 + 9x$
6) $12x^7 - 7x^3$
7) $-2x^6 + 6x^2 + 9x$
8) $-9x^4 - 5x^3 + x$
9) $8x^2 - 21x + 34$
10) $11x^4 - 7x + 8$
11) $-13x^4 + 25x^3 + 45x$
12) $-2x^3 + 9x^2 + 17$
13) $8x^3 + 18x^2 - 8x$
14) $-10x^5 - 9x^4 - 4x^2$
15) $-8x^4 + 7x^2 - 41$
16) $-7x^5 + 3x^3 + 8x^2 - 12$
17) $-9x^5 - 8x^4 + 4x^2 + 12$
18) $-2x^5 - 9x^2 - x$
19) $6x^5 + 7x^4 - 8x^2$
20) $3x^8 - 15x^4 + 11x^2$
21) $-16x^5 + 17x^4 - 9x^2$
22) $-3x^5 + 37x^3 + 5x^2$
23) $-18x^3 + 6x^2 + 15x$
24) $12x^7 + 24x^4$
25) $6x^3 + 48x^2 + 24x$
26) $32x^4 - 16x^2 + 24x$
27) $14x^4 - 14x^3 + 14x$
28) $25x^6 + 20x^5 - 5x$
29) $52x^5 + 4x^4 + 2x^2$
30) $-24x^5 + 42x^3 + 18x^2$

Simplifying Polynomials

1) $3x - 36$
2) $10x^2 - 20x$
3) $35x^2 - 7x$
4) $18x^2 + 12x$
5) $10x^2 - 35x$
6) $9x^2 + 72x$
7) $3x^2 - 17x + 24$
8) $3x^2 - 23x - 36$
9) $x^2 - 13x + 40$
10) $9x^2 - 16$
11) $25x^2 - 50x + 16$
12) $-6x^4 + 14x^2$
13) $7x^3 - 2x^2 + 5x + 10$
14) $-5x^3 + 2x^2 + 8x$
15) $4x^5 - 8x^2 + 15x$
16) $7x^5 + 11x^4 - 4x^2$
17) $-2x^4 + 8x^3 - 14x^2 + 5x$
18) $-10x^3 + x^2 + 31$
19) $2x^3 + x^2 - 4x$
20) $-9x^3 - 4x^2 + 15$
21) $-2x^5 + 2x^4 - 18x^2$
22) $4x^3 - 5$
23) $-12x^5 - 12x^3$
24) -80

ACT Math Workbook

25) $-9x^3 + 2x^2 - 2x + 14$
26) $-4x^3 - 2x^2 + 14$
27) $12x^2 - 7x + 5$
28) $4x^4 - 11x^3 - 5x^2$

29) $6x^2 + 4x - 8$
30) $14x^4 - 5x^2 + 5$
31) $-8x^5 - 36$
32) $10x^3 - 2x$

Adding and Subtracting Polynomials

1) $3x^2 - 1$
2) $4x^2 - 2$
3) $-4x^3 + 4x^2 - 15$
4) $3x^3 - x$
5) $-4x^3 + 14x - 7$
6) $8x^2$
7) $14x^3 - 6$
8) $x^2 - 5$
9) $4x^2 + 4x$

10) $15x$
11) $15x^4 - 5x$
12) $-8x^4 - 5x$
13) $-10x^5 + 7x^3 + 2x$
14) $5x^5 + 5x^3 - 7$
15) $4x^4 + 15x^2$
16) $-12x^2 + 3x$
17) $-21x^4 - x$
18) $-5x^4 + 3x^3 + 4x$

19) $13x^3 - 6x^2 - 7$
20) $14x^5 + 6x^4$
21) $15x^4 + 22x^3 + 8x^2$
22) $9x^2 - 10x$
23) $56x^3 + 5x$
24) $2x^5 - 11x^3 - 2x^2 - 3x$
25) $8x^5 + 18x^3 - 10x$
26) $29x^4 + 11x^3 + 2x^2$

Multiplying Monomials

1) $-7u^7$
2) $36p^{10}$
3) $20xy^3z^5$
4) $16u^7t^3$
5) $10a^5b^3$
6) $-20a^6b^3$
7) $20x^3y^6$
8) $-8p^3q^6$
9) $24s^6t^8$
10) $-49x^9y^4$

11) $15xy^7z^4$
12) $30x^4y^2$
13) $-42p^4q^5$
14) $25s^5t^8$
15) $-24p^8$
16) $-36p^3q^9r^4$
17) $28a^9b$
18) $-36u^{11}v^8$
19) $-27u^6$
20) $-24x^6y^{10}$

21) $-24y^7z^4$
22) $36a^4b^2c^8$
23) $27p^7q^6$
24) $-16u^{16}v^8$
25) $-15y^8z^5$
26) $-60p^5q^6r^4$
27) $9a^3b^6c^6$
28) $63x^{10}y^8z^7$

Multiplying and Dividing Monomials

1) $7x^5$
2) $20x^5$
3) $6x^6$
4) $40x^{11}$

5) $24x^9$
6) $32x^7y$
7) $18x^7y^4$
8) $-10x^5y^7$

9) $32x^6y^5$
10) $-10x^9y^4$
11) $18x^7y^7$
12) $6x^8y^9$

ACT Math Workbook

13) $56x^8y^{18}$
14) $42x^{11}y^{10}$
15) $40x^{12}y^{18}$
16) $-12x^8y^8$
17) $9x^3y^2$

18) xy^2
19) $9x^3y^3$
20) $9xy$
21) $4x^5y^3$
22) $21x^5y$

23) $8x^9y^3$
24) $5x^{-1}y^3$
25) $5x$
26) $5x^{14}y^4$
27) $-4x^2$

Multiplying a Polynomial and a Monomial

1) $2x^2 + 8x$
2) $-3x + 24$
3) $15x^2 + 20x$
4) $-2x^2 + 5x$
5) $21x^2 - 21x$
6) $6x - 15y$
7) $42x^2 - 18x$
8) $12x^2 + 5xy$
9) $5x^2 + 30xy$
10) $44x^2 + 55xy$
11) $32x^2 + 16x$
12) $12 - 180xy$
13) $45x^2 - 27xy$
14) $40x^2 - 16xy + 40x$
15) $18x^3 + 63xy^2$
16) $72x^2 + 48xy$

17) $6x^5 - 2y^5$
18) $-4x^3y + 8xy$
19) $-6x^3 + 9xy - 27$
20) $2x^2 - 4xy - 8$
21) $28x^4 - 7x^2y + 14x^2$
22) $18x^4 + 18x^2 - 63x^2y$
23) $6x^2 + 18xy - 48y^2$
24) $35x^4 - 5x^2 + 40x$
25) $7x^{24} - 28x - 42$
26) $-3x^5 + 4x^3 + 7x^2$
27) $2x^5 - 5x^3 + 10x^2$
28) $12x^7 - 8x^5 + 32x^4$
29) $5x^6 - 25x^3y + 10x^2y^3$
30) $28x^6 - 8x^3 + 44x^2$
31) $21x^6 + 35x^4 - 49x^3$
32) $4x^3 - 32x^2y + 28xy^3$

Multiplying Binomials

1) $x^2 + 6x + 5$
2) $x^2 + 4x - 21$
3) $x^2 - 10x + 9$
4) $x^2 + 11x + 24$
5) $x^2 - 15x + 44$
6) $x^2 + 11x + 30$
7) $x^2 - x - 56$

8) $x^2 - 5x + 6$
9) $x^2 + 19x + 88$
10) $x^2 + 2x + 15$
11) $x^2 + 16x + 64$
12) $x^2 + 9x + 14$
13) $x^2 - 5x - 36$
14) $x^2 - 100$

WWW.MathNotion.Com

15) $x^2 + 26x + 48$
16) $x^2 + 22x + 117$
17) $x^2 - 49$
18) $x^2 - 3x - 10$
19) $3x^2 + 19x + 20$
20) $5x^2 - 38x - 16$
21) $4x^2 - 27x - 81$
22) $6x^2 - 25x + 14$
23) $x^2 + 7x - 44$

24) $10x^2 + 8x - 24$
25) $4x^2 + 19x - 63$
26) $16x^2 + 6x - 10$
27) $21x^2 + 75x + 36$
28) $24x^2 - 8x - 32$
29) $20x^2 - 7x - 40$
30) $64x^2 + 24x - 4$
31) $27x^2 - 42x - 24$
32) $16x^4 - 144$

Factoring Trinomials

1) $(x + 5)(x + 7)$
2) $(x - 2)(x - 6)$
3) $(x + 1)(x + 10)$
4) $(x - 9)(x - 3)$
5) $(x - 1)(x - 15)$
6) $(x - 5)(x - 8)$
7) $(x + 4)(x + 11)$
8) $(x + 9)(x - 8)$

9) $(x - 9)(x + 9)$
10) $(x - 7)(x - 10)$
11) $(x - 4)(x + 12)$
12) $(x - 8)(x + 13)$
13) $(x + 2)(x - 9)$
14) $(x + 11)(x + 11)$
15) $(3x + 9)(x - 4)$
16) $(x - 15)(2x - 5)$

17) $(7x - 5)(2x + 3)$
18) $(2x - 5)(4x + 4)$
19) $(3x + 2)(5x + 2)$
20) $(6x - 1)(4x + 1)$
21) $(x + 5)$
22) $(3x - 2)$
23) $(2x - 7)$

Operations with Polynomials

1) $8x + 2$
2) $10x + 35$
3) $24x - 20$
4) $-28x + 32$
5) $24x^3 + 12x^2$
6) $12x^3 - 54x^2$
7) $-5x^4 + 20x^3$
8) $-20x^5 + 45x^4$

9) $6x^2 + 42x - 18$
10) $12x^2 - 8x + 24$
11) $27x^2 + 72x + 18$
12) $7x^3 + 35x^2 + 21x$
13) $14x^2 - 31x - 10$
14) $24x^2 - 49x - 40$
15) $24x^2 + 8x - 2$
16) $25x^2 + 25x - 36$

17) $(4x + y)$
18) $8x^2 - 22x - 6$
19) $16x^2 + 24x + 9$
20) $40x$
21) $x^3 + 6x^2 + 12x + 8$
22) $(8x + 2y)$

Chapter 9:
Complex Numbers

Topics that you will practice in this chapter:

- ✓ Adding and Subtracting Complex Numbers
- ✓ Multiplying and Dividing Complex Numbers
- ✓ Graphing Complex Numbers
- ✓ Rationalizing Imaginary Denominators

Mathematics is a hard thing to love. It has the unfortunate habit, like a rude dog, of turning its most unfavorable side towards you when you first make contact with it. — *David Whiteland*

Adding and Subtracting Complex Numbers

✎ **Simplify.**

1) $(8i) - (4i) =$

2) $(5i) + (2i) =$

3) $(2i) + (7i) =$

4) $(-6i) - (i) =$

5) $(12i) + (4i) =$

6) $(4i) - (-4i) =$

7) $(-4i) + (-5i) =$

8) $(13i) - (6i) =$

9) $(-21i) - (7i) =$

10) $(-4i) + (2 + 8i) =$

11) $(8 - 4i) + (-6i) =$

12) $(-3i) + (9 + 12i) =$

13) $5 + (9 - 2i) =$

14) $(10i) - (-6 + 2i) =$

15) $(3 + 9i) - (-4i) =$

16) $(7 + 8i) + (-5i) =$

17) $(5i) - (-3i + 4) =$

18) $(6i + 2) + (-2i) =$

19) $(12) - (18 + 3i) =$

20) $(7 + 3i) + (6 + 2i) =$

21) $(4 - 9i) + (3 + 8i) =$

22) $(7 + 3i) + (10 + 12i) =$

23) $(-5 + 5i) - (-5 - 7i) =$

24) $(-8 + 12i) - (-9 + 8i) =$

25) $(-18 + 3i) - (-3 - 12i) =$

26) $(-13 - 4i) + (9 + 12i) =$

27) $(-15 - 2i) - (-14 - 6i) =$

28) $-4 + (8i) + (-14 + 7i) =$

29) $19 - (8i) + (2 - 5i) =$

30) $-3 + (-4 - 8i) - 9 =$

31) $(-24i) + (5 - 8i) + 12 =$

32) $(-11i) - (15 - 12i) + 9i =$

Multiplying and Dividing Complex Numbers

✎ **Simplify.**

1) $(5i)(-3i) =$

2) $(-7i)(2i) =$

3) $(3i)(3i)(-3i) =$

4) $(6i)(-6i) =$

5) $(-7-6i)(7+6i) =$

6) $(4-2i)^2 =$

7) $(5-2i)(4-2i) =$

8) $(5+2i)^2 =$

9) $(7i)(-3i)(9-2i) =$

10) $(2-8i)(6-8i) =$

11) $(-9+3i)(1+4i) =$

12) $(7-8i)(9-3i) =$

13) $5(3i) - (5i)(-4+2i) =$

14) $\dfrac{5}{-25i} =$

15) $\dfrac{12-9i}{-3i} =$

16) $\dfrac{4+9i}{i} =$

17) $\dfrac{20i}{-6+2i} =$

18) $\dfrac{-4-11i}{2i} =$

19) $\dfrac{7i}{3-i} =$

20) $\dfrac{4-9i}{12-5i} =$

21) $\dfrac{8-3i}{-4-4i} =$

22) $\dfrac{-9-5i}{-3-i} =$

23) $\dfrac{-4+i}{-6-5i} =$

24) $\dfrac{-6-7i}{-3+4i} =$

25) $\dfrac{8+11i}{5-5i} =$

Graphing Complex Numbers

✎ **Identify each complex number graphed.**

1)

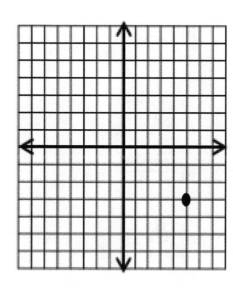

2)

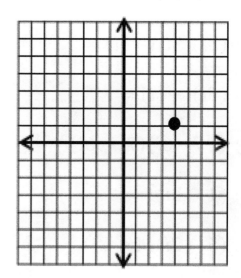

3)

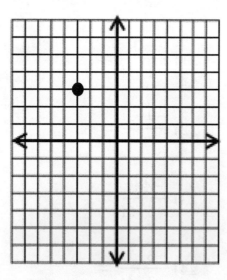

4)

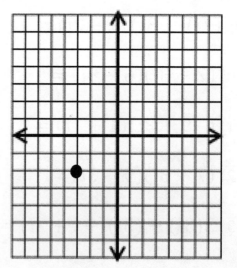

Rationalizing Imaginary Denominators

✍ **Simplify.**

1) $\dfrac{-8}{-8i} =$

2) $\dfrac{-3}{-21i} =$

3) $\dfrac{-14}{-28i} =$

4) $\dfrac{-30}{-5i} =$

5) $\dfrac{9}{2i} =$

6) $\dfrac{27}{-9i} =$

7) $\dfrac{45}{-20i} =$

8) $\dfrac{-26}{8i} =$

9) $\dfrac{6x}{3yi} =$

10) $\dfrac{9-9i}{-3i} =$

11) $\dfrac{4-9i}{-i} =$

12) $\dfrac{12+4i}{3i} =$

13) $\dfrac{8i}{-1+4i} =$

14) $\dfrac{8i}{-2+6i} =$

15) $\dfrac{-15-3i}{-4+4i} =$

16) $\dfrac{-5-9i}{3+4i} =$

17) $\dfrac{-11-4i}{5-2i} =$

18) $\dfrac{-4+6i}{-3i} =$

19) $\dfrac{9+5i}{4i} =$

20) $\dfrac{-5-3i}{7-2i} =$

21) $\dfrac{-8+i}{-3i} =$

22) $\dfrac{9+i}{-5-2i} =$

23) $\dfrac{-9-5i}{-7-2i} =$

24) $\dfrac{9i-5}{-3-6i} =$

Answers of Worksheets – Chapter 9

Adding and Subtracting Complex Numbers

1) $4i$
2) $7i$
3) $9i$
4) $-7i$
5) $16i$
6) $8i$
7) $-9i$
8) $7i$
9) $-28i$
10) $2+4i$
11) $8-10i$
12) $9+9i$
13) $14-2i$
14) $6+8i$
15) $3+13i$
16) $7+3i$
17) $-4+8i$
18) $2+4i$
19) $-6-3i$
20) $13+5i$
21) $7-i$
22) $17+15i$
23) $12i$
24) $1+4i$
25) $-15+15i$
26) $-4+8i$
27) $-1+4i$
28) $-18+15i$
29) $21-13i$
30) $-16-8i$
31) $17-32i$
32) $-15+10i$

Multiplying and Dividing Complex Numbers

1) 15
2) 14
3) $27i$
4) 36
5) $-13-84i$
6) $-16i+12$
7) $16-18i$
8) $21+20i$
9) $189-42i$
10) $-52-64i$
11) $-21-33i$
12) $39-93i$
13) $10+35i$
14) $\frac{i}{5}$
15) $3+4i$
16) $9-4i$
17) $1-3i$
18) $\frac{11}{2}-2i$
19) $-\frac{7}{10}+\frac{21}{10}i$
20) $\frac{93}{169}-\frac{88}{169}i$
21) $-\frac{5}{8}+\frac{11}{8}i$
22) $\frac{16}{5}+\frac{3}{5}i$
23) $\frac{19}{61}-\frac{26}{61}i$
24) $-\frac{2}{5}+\frac{9}{5}i$
25) $-\frac{3}{10}+\frac{19}{10}i$

Graphing Complex Numbers

1) $5-3i$
2) $4+i$
3) $-3+3i$
4) $-3-2i$

Rationalizing Imaginary Denominators

1) $-i$
2) $-\frac{1}{7}i$
3) $\frac{-1}{2}i$
4) $-6i$
5) $-\frac{9}{2}i$
6) $3i$
7) $\frac{9}{4}i$
8) $\frac{13}{4}i$
9) $-\frac{2x}{y}i$
10) $3+3i$
11) $9+4i$
12) $-\frac{4}{3}+4i$
13) $\frac{32}{17}-\frac{8}{17}i$
14) $\frac{6}{5}-\frac{2}{5}i$
15) $\frac{3}{2}+\frac{9}{4}i$
16) $-\frac{51}{25}-\frac{7}{25}i$
17) $-\frac{47}{29}-\frac{42}{29}i$
18) $-2-\frac{4}{3}i$
19) $-\frac{5}{4}+\frac{9}{4}i$
20) $-\frac{29}{53}-\frac{31}{53}i$
21) $-\frac{1}{3}-\frac{8}{3}i$
22) $-\frac{47}{29}+\frac{13}{29}i$
23) $\frac{73}{53}+\frac{17}{53}i$
24) $-\frac{13}{15}-\frac{19}{15}i$

Chapter 10: Functions Operations and Quadratic

Topics that you will practice in this chapter:

- ✓ Graphing Quadratic Functions
- ✓ Solving Quadratic Equations
- ✓ Use the Quadratic Formula and the Discriminant
- ✓ Solve Quadratic Inequalities
- ✓ Evaluating Function
- ✓ Adding and Subtracting Functions
- ✓ Multiplying and Dividing Functions
- ✓ Composition of Functions

It's fine to work on any problem, so long as it generates interesting mathematics along the way – even if you don't solve it at the end of the day." – Andrew Wiles

ACT Math Workbook

Evaluating Function

Write each of following in function notation.

1) $h = -8x + 9$

2) $k = 5a - 21$

3) $d = 14t$

4) $y = \frac{3}{17}x - \frac{9}{17}$

5) $m = 18n - 94$

6) $c = p^2 - 7p + 15$

Evaluate each function.

7) $f(x) = 6x - 7$, find $f(-3)$

8) $g(x) = \frac{1}{10}x + 6$, find $f(5)$

9) $h(x) = -2x + 15$, find $f(8)$

10) $f(x) = -3x + 8$, find $f(-2)$

11) $f(a) = 12a - 9$, find $f(0)$

12) $h(x) = 18 - 5x$, find $f(-4)$

13) $g(n) = 7n - 5$, find $f(5)$

14) $f(x) = -9x - 2$, find $f(3)$

15) $k(n) = -12 + 4.5n$, find $f(2)$

16) $f(x) = -1.5x + 2.5$, find $f(-6)$

17) $g(n) = \frac{16n-8}{6n}$, find $g(2)$

18) $g(n) = \sqrt{5n} - 2$, find $g(5)$

19) $h(x) = x^{-1} - 6$, find $h(\frac{1}{9})$

20) $h(n) = n^{-3} + 4$, find $h(\frac{1}{2})$

21) $h(n) = n^2 - 5$, find $h(\frac{4}{5})$

22) $h(n) = n^3 - 8$, find $h(-\frac{1}{3})$

23) $h(n) = 4n^2 - 42$, find $h(-4)$

24) $h(n) = -5n^2 - 9n$, find $h(7)$

25) $g(n) = \sqrt{4n^2} - \sqrt{5n}$, find $g(5)$

26) $h(a) = \frac{-15a+7}{3a}$, find $h(-b)$

27) $k(a) = 8a - 9$, find $k(a-3)$

28) $h(x) = \frac{1}{6}x + 7$, find $h(-12x)$

29) $h(x) = 8x^2 + 10$, find $h(\frac{x}{2})$

30) $h(x) = x^4 - 8$, find $h(-2x)$

Adding and Subtracting Functions

✏️ **Perform the indicated operation.**

1) $f(x) = 2x + 3$

   $g(x) = x + 4$

   Find $(f - g)(2)$

2) $g(a) = -3a - 8$

   $f(a) = -4a - 12$

   Find $(g - f)(-2)$

3) $h(t) = 7t + 5$

   $g(t) = 3t + 11$

   Find $(h - g)(t)$

4) $g(a) = -5a - 3$

   $f(a) = 3a^2 + 4$

   Find $(g - f)(x)$

5) $g(x) = \frac{2}{7}x - 10$

   $h(x) = \frac{5}{7}x + 10$

   Find $g(14) - h(14)$

6) $h(3) = \sqrt{7x} - 2$

   $g(x) = \sqrt{7x} + 2$

   Find $(h + g)(7)$

7) $f(x) = x^{-3}$

   $g(x) = x^2 + \frac{4}{x}$

   Find $(f - g)(-1)$

8) $h(n) = n^2 + 8$

   $g(n) = -n + 5$

   Find $(h - g)(a)$

9) $g(x) = -2x^2 - 3 - x$

   $f(x) = 7 + x$

   Find $(g - f)(2x)$

10) $g(t) = 4t - 9$

    $f(t) = -t^2 + 5$

    Find $(g + f)(-z)$

11) $f(x) = 3x + 9$

    $g(x) = -4x^2 + 2x$

    Find $(f - g)(-x^2)$

12) $f(x) = -9x^3 - 4x$

    $g(x) = 4x + 12$

    Find $(f + g)(3x^2)$

Multiplying and Dividing Functions

✏️ **Perform the indicated operation.**

1) $g(x) = -2x - 5$

   $f(x) = 3x + 4$

   Find $(g.f)(2)$

2) $f(x) = 3x$

   $h(x) = -2x + 5$

   Find $(f.h)(-3)$

3) $g(a) = 5a - 3$

   $h(a) = a - 7$

   Find $(g.h)(-3)$

4) $f(x) = x - 4$

   $h(x) = 4x - 3$

   Find $(\frac{f}{h})(4)$

5) $f(x) = 9a^2$

   $g(x) = 5 + 4a$

   Find $(\frac{f}{g})(3)$

6) $g(a) = \sqrt{5a} + 7$

   $f(a) = (-a)^2 + 3$

   Find $(\frac{g}{f})(5)$

7) $g(t) = t^2 + 5$

   $h(t) = 2t - 5$

   Find $(g.h)(-3)$

8) $g(n) = n^2 + 2n - 4$

   $h(n) = -n + 6$

   Find $(g.h)(1)$

9) $g(a) = (a - 7)^3$

   $f(a) = a^2 + 8$

   Find $(\frac{g}{f})(7)$

10) $g(x) = -x^2 + \frac{4}{5}x + 10$

    $f(x) = x^2 - 3$

    Find $(\frac{g}{f})(5)$

11) $f(x) = x^3 - 3x^2 + 9$

    $g(x) = x - 4$

    Find $(f.g)(x)$

12) $f(x) = 3x - 5$

    $g(x) = x^2 - 4x$

    Find $(f.g)(x^2)$

Composition of Functions

Using $f(x) = 2x - 5$ and $g(x) = -2x$, find:

1) $f(g(0)) =$

2) $f(g(-1)) =$

3) $g(f(1)) =$

4) $g(f(3)) =$

5) $f(g(-2)) =$

6) $g(f(5)) =$

Using $f(x) = -\frac{1}{4}x + \frac{3}{4}$ and $g(x) = x^2$, find:

7) $g(f(4)) =$

8) $g(f(3)) =$

9) $g(g(2)) =$

10) $f(f(1)) =$

11) $g(f(-1)) =$

12) $g(f(7)) =$

Using $f(x) = -5x + 2$ and $g(x) = x + 3$, find:

13) $g(f(0)) =$

14) $f(f(2)) =$

15) $f(g(3)) =$

16) $f(g(-3)) =$

17) $g(f(-5)) =$

18) $f(f(x)) =$

Using $f(x) = \sqrt{x + 9}$ and $g(x) = x - 9$, find:

19) $f(g(9)) =$

20) $g(f(-8)) =$

21) $f(g(18)) =$

22) $f(f(-5)) =$

23) $g(f(7)) =$

24) $g(g(8)) =$

Quadratic Equation

✎ Multiply.

1) $(x - 2)(x + 8) = $ _____

2) $(x + 1)(x + 9) = $ _____

3) $(x - 5)(x + 6) = $ _____

4) $(x + 7)(x - 3) = $ _____

5) $(x - 9)(x - 8) = $ _____

6) $(4x + 2)(x - 4) = $ _____

7) $(3x - 6)(x + 4) = $ _____

8) $(x - 9)(2x + 7) = $ _____

9) $(5x + 3)(x - 4) = $ _____

10) $(4x + 2)(3x - 3) = $ _____

✎ Factor each expression.

11) $x^2 - 4x - 21 = $ _____

12) $x^2 + 14x + 45 = $ _____

13) $x^2 - 5x - 24 = $ _____

14) $x^2 - 7x + 6 = $ _____

15) $x^2 + 14x + 33 = $ _____

16) $4x^2 + 38x + 18 = $ _____

17) $5x^2 + 18x - 8 = $ _____

18) $2x^2 + 2x - 40 = $ _____

19) $2x^2 + 22x + 56 = $ _____

20) $12x^2 - 148x + 360 = $ _____

✎ Calculate each equation.

21) $(x + 6)(x - 9) = 0$

22) $(x + 1)(x + 11) = 0$

23) $(3x + 9)(x + 3) = 0$

24) $(5x - 5)(6x + 12) = 0$

25) $x^2 - 12x + 30 = 6$

26) $x^2 + 6x + 14 = 5$

27) $x^2 + \frac{9}{2}x + 7 = 5$

28) $x^2 + 2x - 25 = 10$

29) $2x^2 + 12x - 54 = 0$

30) $x^2 - 11x = 12$

Solving Quadratic Equations

✎ **Solve each equation by factoring or using the quadratic formula.**

1) $(x+5)(x-2) = 0$

2) $(x+8)(x+2) = 0$

3) $(x-9)(x+5) = 0$

4) $(x-3)(x-1) = 0$

5) $(x+9)(x+4) = 0$

6) $(2x+5)(x+9) = 0$

7) $(9x+8)(3x+9) = 0$

8) $(4x+2)(x+5) = 0$

9) $(x+2)(2x+9) = 0$

10) $(12x+3)(2x+9) = 0$

11) $2x^2 = 16x$

12) $x^2 - 16 = 0$

13) $2x^2 + 48 = 22x$

14) $-x^2 - 20 = 9x$

15) $x^2 + 8x = 33$

16) $2x^2 + 12x = 80$

17) $x^2 + 14x = -48$

18) $x^2 + 15x = -54$

19) $x^2 + 15x = -36$

20) $x^2 + 2x - 40 = 5x$

21) $x^2 + 16x = -63$

22) $x^2 - 18x = -81$

23) $10x^2 = 7x - 1$

24) $7x^2 - 5x + 8 = 8$

25) $8x^2 + 27 = 33x$

26) $5x^2 - 26x = -24$

27) $3x^2 + 6 = -19x$

28) $x^2 + 22x = -117$

29) $x^2 + 3x - 58 = 30$

30) $5x^2 + 20x - 200 = 25$

31) $3x^2 - 33x + 84 = 0$

32) $6x^2 - 31x + 30 = 15 - 10x^2$

Quadratic Formula and the Discriminant

✍ **Find the value of the discriminant of each quadratic equation.**

1) $3x(x-9) = 0$

2) $2x^2 + 9x - 4 = 0$

3) $x^2 + 9x + 5 = 0$

4) $4x^2 - 4x + 7 = 0$

5) $x^2 + 7x - 6 = 0$

6) $4x^2 + 5x - 13 = 0$

7) $3x^2 + 7x + 11 = 0$

8) $x^2 - 4x - 12 = 0$

9) $5x^2 + 9x + 8 = 0$

10) $x^2 + 3x - 7 = 0$

11) $6x^2 + 7x - 13 = 0$

12) $-8x^2 - 11x + 9 = 0$

13) $-9x^2 - 13x + 7 = 0$

14) $-6x^2 - 7x - 9 = 0$

15) $14x^2 - 8x - 15 = 0$

16) $-9x^2 - 5x + 10 = 0$

17) $8x^2 + 9x - 14 = 0$

18) $7x^2 - 15x = 0$

19) $3x^2 - 7x + 9 = 0$

20) $7x^2 + 4x + 16 = 0$

✍ **Find the discriminant of each quadratic equation then state the number of real and imaginary solutions.**

21) $-x^2 - 4 = 4x$

22) $20x^2 = 20x - 5$

23) $-11x^2 - 11x = 22$

24) $19x^2 - 4x + 1 = 15x^2$

25) $-8x^2 = -6x + 6$

26) $2x^2 + 4x + 4 = 2$

27) $6x^2 - 2x - 9 = -12$

28) $-14x^2 - 56x - 64 = -8$

Quadratic Inequalities

✎ **Solve each quadratic inequality.**

1) $x^2 - 64 < 0$

2) $-x^2 - 6x - 8 > 0$

3) $x^2 + 6x + 8 < 0$

4) $4x^2 + 28x + 40 > 0$

5) $5x^2 - 5x - 10 \geq 0$

6) $3x^2 > -12x - 27$

7) $4x^2 + 10x + 28 \leq 0$

8) $3x^2 - 9x - 30 \leq 0$

9) $5x^2 - 35x + 60 \geq 0$

10) $x^2 + 7x + 10 < 0$

11) $2x^2 + 16x - 130 > 0$

12) $8x^2 - 24x + 18 > 0$

13) $2x^2 - 32x + 136 \leq 0$

14) $x^2 - 14x + 49 \leq 0$

15) $2x^2 - 30x + 112 \geq 0$

16) $2x^2 + 16x + 32 \leq 0$

17) $x^2 - 121 \leq 0$

18) $2x^2 - 22x + 60 \geq 0$

19) $8x^2 + 10x + 18 \leq 0$

20) $4x^2 - 2x - 24 > 2x^2$

21) $4x^2 - 16x + 16 < 0$

22) $15x^2 - 6x \geq 14x^2 - 5$

23) $6x^2 - 24 > 4x^2 + 2x$

24) $3x^2 - x \geq 3x^2 - 4x + 6$

25) $2x^2 + 2x - 8 > x^2$

26) $4x^2 + 20x - 11 < 0$

27) $-2x^2 + 30x - 114 \geq 0$

28) $-8x^2 + 6x - 1 \leq 0$

29) $x^2 + 7x + 10 < 0$

30) $36x^2 + 46x + 10 \leq 0$

31) $5x^2 + 5x - 60 \geq 0$

32) $3x^2 + 4x \leq 2x^2 + 2x - 10$

Graphing Quadratic Functions

✎ **Sketch the graph of each function. Identify the vertex and axis of symmetry.**

1) $y = (x + 3)^2 + 5$

2) $y = (x - 3)^2 - 1$

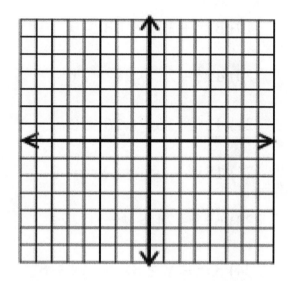
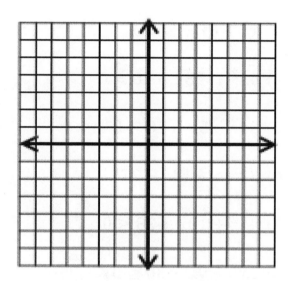

3) $y = 6 - (-x + 2)^2$

4) $y = -3x^2 - 6x + 9$

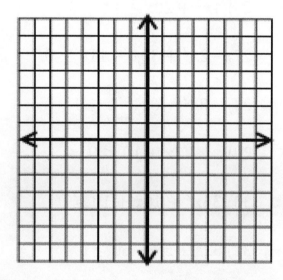
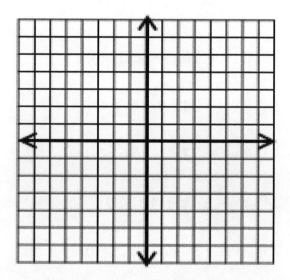

Domain and Range of Radical Functions

✎ **Identify the domain and range of each function.**

1) $y = \sqrt{x+6} - 9$

2) $y = \sqrt[3]{5x-4} - 12$

3) $y = \sqrt{3x-9} + 7$

4) $y = \sqrt[3]{(8x+11)} - 9$

5) $y = 2\sqrt{6x+30} + 14$

6) $y = \sqrt[3]{(9x-15)} - 17$

7) $y = 2\sqrt{9x^2+18} + 7$

8) $y = \sqrt[3]{(8x^2-5)} - 13$

9) $y = \sqrt{2x^3+16} - 9$

10) $y = \sqrt[3]{(14x+3)} - 5x$

11) $y = 3\sqrt{-3(12x+24)} + 7$

12) $y = \sqrt[5]{(11x^2-17)} - 21$

13) $y = 4\sqrt{x-9} - 27$

14) $y = \sqrt[3]{5x+1} - 3$

✎ **Sketch the graph of each function.**

15) $y = -3\sqrt{x} + 4$

16) $y = 6\sqrt{x} - 8$

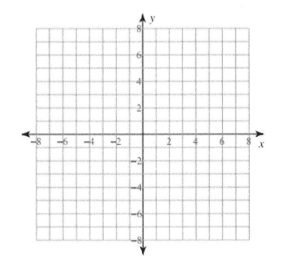
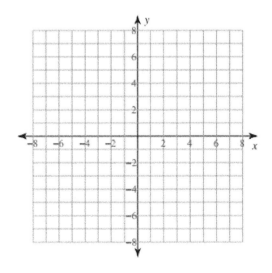

Solving Radical Equations

✎ **Solve each equation. Remember to check for extraneous solutions.**

1) $\sqrt{a} = 9$

2) $\sqrt{v} = 4$

3) $\sqrt{r} = 7$

4) $4 = 16\sqrt{x}$

5) $\sqrt{x+3} = 12$

6) $2 = \sqrt{x-8}$

7) $7 = \sqrt{r-6}$

8) $\sqrt{x-4} = 9$

9) $15 = \sqrt{x-6}$

10) $\sqrt{m+8} = 11$

11) $5\sqrt{3a} = 75$

12) $2\sqrt{10x} = 30$

13) $4 = \sqrt{3x-10}$

14) $\sqrt{150-3x} = 2$

15) $\sqrt{r+4} - 8 = 8$

16) $-18 = -3\sqrt{r+2}$

17) $60 = 6\sqrt{49v}$

18) $3 = \sqrt{50-x}$

19) $\sqrt{90-5a} = 6$

20) $\sqrt{-3n+33} = 3$

21) $\sqrt{21r-18} = 4r$

22) $\sqrt{-14+6x} = 7x$

23) $\sqrt{4x+15} = \sqrt{2x+11}$

24) $\sqrt{8v} = \sqrt{10v-14}$

25) $\sqrt{16-3x} = \sqrt{3x-8}$

26) $\sqrt{5m+12} = \sqrt{7m+12}$

27) $\sqrt{8r+15} = \sqrt{-13-5r}$

28) $\sqrt{4k+6} = \sqrt{2-8k}$

29) $-60\sqrt{x-16} = -120$

30) $\sqrt{20-x} = \sqrt{\dfrac{x}{4}}$

Answers of Worksheets – Chapter 10

Evaluating Function

1) $h(x) = -8x + 9$
2) $k(a) = 5a - 21$
3) $d(t) = 14t$
4) $y(x) = \frac{3}{17}x - \frac{9}{17}$
5) $m(n) = 18n - 94$
6) $c(p) = p^2 - 7p + 15$
7) -25
8) 6.5
9) -1
10) 14
11) -9
12) 38
13) 30
14) -29
15) -3
16) 11.5
17) 2
18) 3
19) 3
20) 12
21) $-\frac{109}{25}$
22) $-\frac{215}{27}$
23) 22
24) -308
25) 5
26) $-\frac{15b+7}{3b}$
27) $8a - 33$
28) $-2x + 7$
29) $2x^2 + 10$
30) $-16x^4 - 8$

Adding and Subtracting Functions

1) 1
2) 2
3) $4t - 6$
4) $-3x^2 - 5x - 7$
5) -26
6) 14
7) 2
8) $a^2 + a + 3$
9) $-8x^2 - 4x - 10$
10) $-z^2 - 4z - 4$
11) $4x^4 - x^2 + 9$
12) $-243x^6 + 12$

Multiplying and Dividing Functions

1) -90
2) -99
3) 180
4) 0
5) $\frac{81}{17}$
6) $\frac{3}{7}$
7) -154
8) -5
9) 0
10) $-\frac{1}{2}$
11) $x^4 - 7x^3 + 12x^2 + 9x - 36$
12) $3x^6 - 17x^4 + 20x^2$

Composition of Functions

1) -5 2) -1 3) 6 4) -2

ACT Math Workbook

5) 3
6) -10
7) $\frac{1}{16}$
8) 0
9) 16
10) $\frac{5}{8}$
11) 1
12) 1
13) 5
14) 42
15) -28
16) 2
17) 30
18) $25x - 8$
19) 3
20) -8
21) $3\sqrt{2}$
22) $\sqrt{11}$
23) -5
24) -10

Quadratic Equations

1) $x^2 + 6x - 16$
2) $x^2 + 10x + 9$
3) $x^2 + x - 30$
4) $x^2 + 4x - 21$
5) $x^2 - 17x + 72$
6) $4x^2 - 14x - 8$
7) $3x^2 + 6x - 24$
8) $2x^2 - 11x - 63$
9) $5x^2 - 17x - 12$
10) $12x^2 - 6x - 6$
11) $(x - 7)(x + 3)$
12) $(x + 5)(x + 9)$
13) $(x - 8)(x + 3)$
14) $(x - 1)(x - 6)$
15) $(x + 3)(x + 11)$
16) $(4x + 2)(x + 9)$
17) $(5x - 2)(x + 4)$
18) $(2x - 8)(x + 5)$
19) $(2x + 8)(x + 7)$
20) $4(x - 9)(3x - 10)$
21) $x = -6, x = 9$
22) $x = -1, x = -11$
23) $x = -3$
24) $x = 1, x = -2$
25) $x = 6$
26) $x = -3$
27) $x = -4, x = -\frac{1}{2}$
28) $x = 5, x = -7$
29) $x = 3, x = -9$
30) $x = -1, x = 12$

Solving quadratic equations

1) $\{-5, 2\}$
2) $\{-8, -2\}$
3) $\{9, -5\}$
4) $\{3, 1\}$
5) $\{-9, -4\}$
6) $\{-\frac{5}{2}, -9\}$
7) $\{-\frac{8}{9}, -3\}$
8) $\{-\frac{1}{2}, -5\}$
9) $\{-2, -\frac{9}{2}\}$
10) $\{-\frac{1}{4}, -\frac{9}{2}\}$
11) $\{8, 0\}$
12) $\{4, -4\}$
13) $\{3, 8\}$
14) $\{-5, -4\}$
15) $\{3, -11\}$
16) $\{4, -10\}$
17) $\{-6, -8\}$
18) $\{-6, -9\}$
19) $\{-3, -12\}$
20) $\{8, -5\}$
21) $\{-7, -9\}$
22) $\{9\}$
23) $\{\frac{1}{5}, \frac{1}{2}\}$
24) $\{\frac{5}{7}, 0\}$
25) $\{\frac{9}{8}, 3\}$
26) $\{\frac{6}{5}, 4\}$
27) $\{-\frac{1}{3}, -6\}$
28) $\{-9, -13\}$
29) $\{8, -11\}$
30) $\{5, -9\}$
31) $\{4, 7\}$
32) $\{\frac{15}{16}, 1\}$

Quadratic formula and the discriminant

WWW.MathNotion.Com

ACT Math Workbook

1) 729
2) 113
3) 61
4) −96
5) 73
6) 233
7) 83
8) 8
9) −79
10) 37
11) 361
12) 409
13) 421
14) 7
15) 904
16) 385
17) 529
18) 225
19) −59
20) −432
21) 0, one real solution
22) 0, one real solution
23) −847, no solution
24) 0, one real solution
25) −156, no solution
26) 0, one real solution
27) −68, no solution
28) 0, one real solution

Solve quadratic inequalities

1) $-8 < x < 8$
2) $-4 < x < -2$
3) $-4 < x < -2$
4) $x < -5 \text{ or } x > -2$
5) $x \leq -1 \text{ or } x \geq 2$
6) all real numbers
7) no solution
8) $-2 \leq x \leq 5$
9) $x \leq 3 \text{ or } x \geq 4$
10) $-5 < x < -2$
11) $x < -13 \text{ or } x > 5$
12) $x < \frac{3}{2} \text{ or } x > \frac{3}{2}$
13) no solution
14) $x = 7$
15) $x \leq 7 \text{ or } x \geq 8$
16) $x = -4$
17) $-11 \leq x \leq 11$
18) $x \leq 5 \text{ or } x \geq 6$
19) no solution
20) $x < -3 \text{ or } x > 4$
21) no solution
22) $x \leq 1 \text{ or } x \geq 5$
23) $x < -3 \text{ or } x > 4$
24) $x \geq 2$
25) $x < -4 \text{ or } x > 2$
26) $-\frac{11}{2} < x < \frac{1}{2}$
27) no solution
28) $x \leq \frac{1}{4} \text{ or } x \geq \frac{1}{2}$
29) $-5 < x < -2$
30) $-1 \leq x \leq -\frac{5}{18}$
31) $x \leq -4 \text{ or } x \geq 3$
32) no solution

ACT Math Workbook

Graphing quadratic functions

1) $(-3, 5), x = -3$

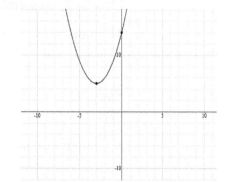

2) $(3, -1), x = 3$

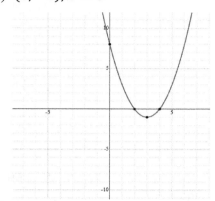

3) $(2, 6), x = 2$

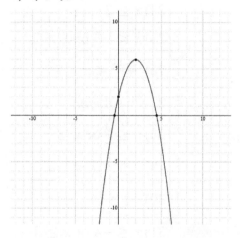

5) $(-1, 12), x = -1$

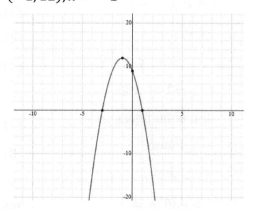

Domain and range of radical functions

1) domain: $x \geq -6$
   range: $y \geq -9$

2) domain: {all real numbers}
   range: {all real numbers}

3) domain: $x \geq 3$
   range: $y \geq 7$

4) domain: {all real numbers}
   range: {all real numbers}

5) domain: $x \geq -5$
   range: $y \geq 14$

6) domain: {all real numbers}
   range: {all real numbers}

7) domain: {all real numbers}
   range: $y \geq 6\sqrt{2} + 7$

8) domain: {all real numbers}
   range: {all real numbers}

9) domain: $x \geq -2$
   range: $y \geq -9$
10) domain: {all real numbers}
    range: {all real numbers}
11) domain: $x \leq -2$
    range: $y \geq 7$
12) domain: {all real numbers}
    range: {all real numbers}
13) domain: $x \geq 9$
    range: $y \geq -27$
14) domain: {all real numbers}
    range: {all real numbers}

15)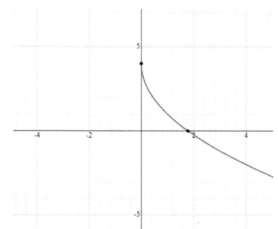

16)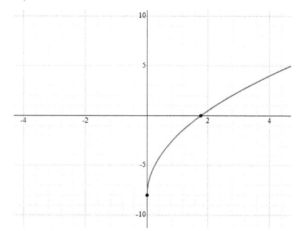

Solving radical equations

1) {81}
2) {16}
3) {49}
4) {$\frac{1}{16}$}
5) {141}
6) {12}
7) {55}
8) {85}
9) {231}
10) {113}
11) {75}
12) {22.5}
13) {$\frac{26}{3}$}
14) $\frac{146}{3}$
15) {252}
16) {34}
17) {$\frac{100}{49}$}
18) {41}
19) {54/5}
20) {8}
21) no solution
22) no solution
23) {−2}
24) {7}
25) {4}
26) {0}
27) no solution
28) {$-\frac{1}{3}$}
29) {20}
30) {16}

ACT Math Workbook

Chapter 11:
Sequences and Series

Topics that you will practice in this chapter:

- ✓ Arithmetic Sequences
- ✓ Geometric Sequences
- ✓ Comparing Arithmetic and Geometric Sequences
- ✓ Finite Geometric Series
- ✓ Infinite Geometric Series

Mathematics is like checkers in being suitable for the young, not too difficult, amusing, and without peril to the state. — *Plato*

Arithmetic Sequences

✎ **Find the next three terms of each arithmetic sequence.**

1) 32, 23, 14, 5, −4, …

2) −91, −63, −35, −7, …

3) 51, 62, 73, 84, 95, …

4) 84, 53, 22, −9, −40, …

✎ **Given the first term and the common difference of an arithmetic sequence find the first five terms and the explicit formula.**

5) $a_1 = 9, d = 12$

6) $a_1 = -10, d = -5$

7) $a_1 = 52, d = 22$

8) $a_1 = 210, d = -102$

✎ **Given a term in an arithmetic sequence and the common difference find the first five terms and the explicit formula.**

9) $a_{51} = -468, d = -12$

10) $a_{31} = 230, d = 6$

11) $a_{62} = -128.2, d = -4.2$

12) $a_{33} = -2{,}352, d = -77$

✎ **Given a term in an arithmetic sequence and the common difference find the recursive formula and the three terms in the sequence after the last one given.**

13) $a_{25} = -156, d = -6$

14) $a_{16} = 111, d = 7.1$

15) $a_{22} = 43, d = 1.8$

16) $a_{14} = -17, d = 0.4$

Geometric Sequences

✎ **Determine if the sequence is geometric. If it is, find the common ratio.**

1) $2, -14, 98, -686, \ldots$

2) $-3, -15, -75, -375, \ldots$

3) $7, 17, 31, 126, \ldots$

4) $-5, -35, -245, -1715, \ldots$

✎ **Given the first term and the common ratio of a geometric sequence find the first five terms and the explicit formula.**

5) $a_1 = 0.7, r = -3$

6) $a_1 = 0.4, r = 5$

✎ **Given the recursive formula for a geometric sequence find the common ratio, the first five terms, and the explicit formula.**

7) $a_n = a_{n-1} \times 6, a_1 = 2$

8) $a_n = a_{n-1} \cdot (-4), a_1 = -6$

9) $a_n = a_{n-1} \cdot 9, a_1 = 0.2$

10) $a_n = a_{n-1} \cdot 3, a_1 = -8$

✎ **Given two terms in a geometric sequence find the 9th term and the recursive formula.**

11) $a_5 = 729$ and $a_6 = -243$

12) $a_6 = -768$ and $a_3 = 12$

Comparing Arithmetic and Geometric Sequences

✍ For each sequence, state if it is arithmetic, geometric, or neither.

1) 5, 10, 15, 20, …

2) 6, 10, 14, 20, …

3) 2, 6, 24, 51, …

4) 1, 8, 18, 28, 36, …

5) 2, 8, 17, 52, 142, …

6) 2, 5, 9, 17, 36, …

7) 0.6, 3, 15, 75, 375, …

8) 4, 20, 100, 500, …

9) −18, −23, −28, −33, −38, …

10) −3, 12, −48, 192, −768, …

11) 8, 18, 26, 39, 50, …

12) 3, 12, 90, 150, 210 …

13) −22, −12, −2, 2, 12, …

14) $a_n = 2 \cdot 7^{n-1}$

15) $a_n = 8 \cdot 4^{n-1}$

16) $a_n = 9 - 5n$

17) $a_n = -110 + 200n$

18) $a_n = 15 + 13n$

19) $a_n = -6 \cdot (-11)^{n-1}$

20) $a_n = 120 + 42n$

21) $a_n = (4n)^4$

22) $a_n = 28 + 6n$

23) $a_n = -(13)^{n-1}$

24) $a_n = -7 \cdot (1.5)^{n-1}$

25) $a_n = \frac{2n+1}{7n}$

26) $a_n = \frac{24-13n}{6n}$

27) $a_n = \frac{8-17n}{2n}$

28) $a_n = \frac{32 - a_{n-1}}{9}$

29) $a_n = -\frac{3}{19} + \frac{2}{7}n$

Finite Geometric Series

Evaluate the related series of each sequence.

1) $-2, 8, -32, 128$

2) $-1, 3, -9, 27, -81$

3) $-1, 4, -16, 64, -256$

4) $1, 8, 64, 512$

5) $-6, -24, -96, -384$

6) $2, -12, 72, -432, 2592$

Evaluate each geometric series described.

7) $1 + 3 + 9 + 27 \ldots, n = 6$ _____

8) $1.5 - 6 + 24 - 96 \ldots, n = 6$ _____

9) $-2 - 6 - 18 - 54 \ldots, n = 7$ _____

10) $0.5 - 3 + 18 - 108 \ldots, n = 6$ _____

11) $2.5 - 10 + 40 - 160 \ldots, n = 8$ _____

12) $-1 + 7 - 49 + 343 \ldots, n = 6$ _____

13) $a_1 = -2, r = 6, n = 5$ _____

14) $a_1 = 3, r = 2, n = 9$ _____

15) $\sum_{n=1}^{5} 4 \cdot (-3)^{n-1}$ _____

16) $\sum_{n=1}^{7} 6 \cdot (-2)^{n-1}$ _____

17) $\sum_{n=1}^{5} 3 \cdot (5)^{n-1}$ _____

18) $\sum_{m=1}^{10} (-2)^{m-1}$ _____

19) $\sum_{m=1}^{4} 8 \times (5)^{m-1}$ _____

20) $\sum_{k=1}^{8} 2 \times (4)^{k-1}$ _____

Infinite Geometric Series

✎ Determine if each geometric series converges or diverges.

1) $a_1 = -1.4, \ r = 6$

2) $a_1 = 5.2, r = 0.3$

3) $a_1 = -6, r = 7.2$

4) $a_1 = 12, r = 0.04$

5) $a_1 = 3, r = 15$

6) $-1, 7, -49, 343, \ldots$

7) $6, -1, \frac{1}{6}, -\frac{1}{36}, \frac{1}{216}, \ldots$

8) $512 + 64 + 8 + 1 \ldots$

9) $-4 + \frac{12}{7} - \frac{36}{49} + \frac{108}{343} \ldots$

10) $\frac{120}{459} - \frac{60}{153} + \frac{30}{51} - \frac{15}{17} \ldots$

✎ Evaluate each infinite geometric series described.

11) $a_1 = 4, r = -\frac{1}{6}$

12) $a_1 = 18, r = -\frac{1}{3}$

13) $a_1 = 16, r = -\frac{1}{7}$

14) $a_1 = 8, r = \frac{1}{3}$

15) $2 + 0.5 + 0.125 + 0.031 + \cdots$

16) $125 - 25 + 5 - 1 \ldots,$

17) $1 - 0.6 + 0.36 - 0.216 \ldots,$

18) $3 + \frac{12}{5} + \frac{48}{25} + \frac{192}{125} \ldots,$

19) $\sum_{k=1}^{\infty} 11^{k-1}$

20) $\sum_{i=1}^{\infty} \left(\frac{2}{5}\right)^{i-1}$

21) $\sum_{k=1}^{\infty} \left(-\frac{3}{7}\right)^{k-1}$

22) $\sum_{n=1}^{\infty} 12\left(\frac{5}{6}\right)^{n-1}$

Answers of Worksheets – Chapter 11

Arithmetic Sequences

1) $-13, -22, -31$

2) $21, 49, 77$

3) $106, 117, 128$

4) $-71, -102, -133$

5) First Five Terms: $9, 21, 33, 45, 57$, Explicit: $a_n = 9 + 12(n-1)$

6) First Five Terms: $-10, -15, -20, -25, -30$, Explicit: $a_n = -10 - 5(n-1)$

7) First Five Terms: $52, 74, 96, 118, 140$, Explicit: $a_n = 52 + 22(n-1)$

8) First Five Terms: $210, 108, 6, -96, -198$, Explicit: $a_n = 210 - 102(n-1)$

9) First Five Terms: $132, 120, 108, 96, 84$, Explicit: $a_n = 132 - 12(n-1)$

10) First Five Terms: $50, 56, 62, 68, 74$, Explicit: $a_n = 50 + 6(n-1)$

11) First Five Terms: $128, 123.8, 119.6, 115.4, 111.2$, Explicit: $a_n = 128 - 4.2(n-1)$

12) First Five Terms: $112, 35, -42, -119, -196$, Explicit: $a_n = 112 - 77(n-1)$

13) Next 3 terms: $-162, -168, -174$, Recursive: $a_n = a_{n-1} - 6, a_1 = -6$

14) Next 3 terms: $118.2, 125.2, 132.3$ Recursive: $a_n = a_{n-1} + 7.1, a_1 = 4.5$

15) Next 3 terms: $44.8, 46.6, 48.4$, Recursive: $a_n = a_{n-1} + 1.8, a_1 = 5.2$

16) Next 3 terms: $-9.8, -9.6, -9.4$, Recursive: $a_n = a_{n-1} + 0.4, a_1 = -22.2$

Geometric Sequences

1) $r = -7$

2) $r = 5$

3) not geometric

4) $r = 7$

5) First Five Terms: $0.7, -2.1, 6.3, -18.9, 56.7$

   Explicit: $a_n = 0.7 \times (-3)^{n-1}$

6) First Five Terms: $0.4, 2, 10, 50, 250$

   Explicit: $a_n = 0.4 \times (5)^{n-1}$

7) Common Ratio: $r = 6$

   First Five Terms: $2, 12, 72, 432, 2{,}592$

   Explicit: $a_n = 2 \cdot (6)^{n-1}$

ACT Math Workbook

8) Common Ratio: $r = -4$

   First Five Terms: $-6, 24, -96, 384, -1,536$

   Explicit: $a_n = -6 \cdot (-4)^{n-1}$

9) Common Ratio: $r = 9$

   First Five Terms: $0.2; 1.8; 16.2; 145.8; 1,312.2; 11,809.8$

   Explicit: $a_n = 0.2 \cdot (9)^{n-1}$

10) Common Ratio: $r = 3$

    First Five Terms: $-8, -24, -72, -216, -648$

    Explicit: $a_n = -8 \cdot (3)^{n-1}$

11) $a_9 = 9$, Recursive: $a_n = a_{n-1} \cdot (\frac{-1}{3})$, $a_1 = 59,049$

12) $a_9 = 49,152$, Recursive: $a_n = a_{n-1} \cdot (-4)$, $a_1 = 0.75$

Comparing Arithmetic and Geometric Sequences

1) Arithmetic
2) Arithmetic
3) Neither
4) Neither
5) Neither
6) Neither
7) Geometric
8) Geometric
9) Arithmetic
10) Geometric
11) Neither
12) Neither
13) Arithmetic
14) Geometric
15) Geometric
16) Arithmetic
17) Arithmetic
18) Arithmetic
19) Geometric
20) Arithmetic
21) Neither
22) Arithmetic
23) Geometric
24) Geometric
25) Neither
26) Neither
27) Neither
28) Neither
29) Arithmetic

Finite Geometric

1) 102
2) -61
3) -205
4) 585
5) -510
6) 2,157
7) 364
8) $-1,228.5$
9) $-2,186$
10) $-3,333.5$
11) $-32,767.5$
12) 14,699
13) $-3,110$
14) 1,533
15) 244
16) 258
17) 2,343
18) -341
19) 1,248
20) 43,680

ACT Math Workbook

Infinite Geometric

1) Diverges
2) Converges
3) Diverges
4) Converges
5) Diverges
6) Diverges
7) Converges
8) Converges
9) Converges
10) Diverges
11) $\frac{24}{7}$
12) $\frac{27}{2}$
13) 14
14) 12
15) $\frac{8}{3}$
16) $\frac{625}{6}$
17) $\frac{5}{8}$
18) 15
19) Infinite
20) $\frac{5}{3}$
21) $\frac{7}{10}$
22) 72

Chapter 12: Logarithms

Topics that you will practice in this chapter:

- ✓ Rewriting Logarithms
- ✓ Evaluating Logarithms
- ✓ Properties of Logarithms
- ✓ Natural Logarithms
- ✓ Exponential Equations Requiring Logarithms
- ✓ Solving Logarithmic Equations

Mathematics is an art of human understanding. — *William Thurston*

Rewriting Logarithms

✎ **Rewrite each equation in exponential form.**

1) $\log_5 25 = 2$

2) $\log_4 256 = 4$

3) $\log_3 81 = 4$

4) $\log_8 64 = 2$

5) $\log_6 216 = 3$

6) $\log_2 16 = 4$

7) $\log_{10} 100 = 2$

8) $\log_3 243 = 5$

9) $\log_5 625 = 4$

10) $\log_2 256 = 8$

11) $\log_3 6,561 = 8$

12) $\log_{11} 121 = 2$

13) $\log_{14} 196 = 2$

14) $\log_{81} 3 = \frac{1}{4}$

15) $\log_{27} 3 = \frac{1}{3}$

16) $\log_{32} 2 = \frac{1}{5}$

17) $\log_{512} 8 = \frac{1}{3}$

18) $\log_2 \frac{1}{8} = -3$

19) $\log_2 \frac{1}{16} = -4$

20) $\log_a \frac{7}{3} = b$

✎ **Rewrite each exponential equation in logarithmic form.**

21) $12^2 = 144$

22) $7^3 = 343$

23) $4^5 = 1,024$

24) $15^2 = 225$

25) $5^4 = 625$

26) $6^4 = 1,296$

27) $2^9 = 512$

28) $5^5 = 3,125$

29) $4^{-6} = \frac{1}{4,096}$

30) $3^{-5} = \frac{1}{243}$

31) $16^{-2} = \frac{1}{256}$

32) $6^{-3} = \frac{1}{216}$

33) $3^{-9} = \frac{1}{19,683}$

34) $21^{-2} = \frac{1}{441}$

Evaluating Logarithms

✎ **Evaluate each logarithm.**

1) $\log_3 2{,}187 =$

2) $\log_2 256 =$

3) $\log_5 125 =$

4) $\log_5 625 =$

5) $\log_3 243 =$

6) $\log_4 1{,}024 =$

7) $\log_8 64 =$

8) $\log_8 \frac{1}{8} =$

9) $\log_6 \frac{1}{36} =$

10) $\log_2 \frac{1}{16} =$

11) $\log_6 \frac{1}{216} =$

12) $\log_3 \frac{1}{256} =$

13) $\log_{18} \frac{1}{324} =$

14) $\log_{256} \frac{1}{4} =$

15) $\log_{512} 8 =$

16) $\log_4 \frac{1}{4{,}096} =$

17) $\log_9 \frac{1}{729} =$

18) $\log_{216} \frac{1}{6} =$

✎ **Circle the points which are on the graph of the given logarithmic functions.**

19) $y = 5\log_8(3x - 4) + 1$ $(6, 5),$ $(4, 6),$ $(4, 8)$

20) $y = 3\log_2(4x) - 6$ $(4, 6),$ $(\frac{1}{4}, 16),$ $(\frac{1}{4}, -6)$

21) $y = -3\log_5(x - 2) + 5$ $(7, -2),$ $(7, 2),$ $(6, -3)$

22) $y = \frac{1}{2}\log_6(6x) + 4$ $(6, 5),$ $(6, \frac{1}{5}),$ $(6, -5)$

23) $y = -\log_9 9(x + 5) + 4$ $(-4, 2),$ $(4, 0),$ $(4, 2)$

24) $y = -\log_8(x - 6) - 4$ $(7, -\frac{1}{4}),$ $(7, -4),$ $(7, -\frac{1}{4})$

25) $y = -3\log_6(x + 3) + 6$ $(3, 3),$ $(-5, -3),$ $(-5, 3)$

Properties of Logarithms

✎ **Expand each logarithm.**

1) $\log(9 \times 4) =$

2) $\log(6 \times 3) =$

3) $\log(2 \times 8) =$

4) $\log\left(\frac{8}{7}\right) =$

5) $\log\left(\frac{9}{5}\right) =$

6) $\log\left(\frac{4}{11}\right)^3 =$

7) $\log(9 \times 4^3) =$

8) $\log\left(\frac{7}{3}\right)^2 =$

9) $\log\left(\frac{5^4}{9}\right) =$

10) $\log(x \times y)^7 =$

11) $\log(x^2 \times y \times z^5) =$

12) $\log\left(\frac{u^8}{v}\right) =$

13) $\log\left(\frac{x}{y^4}\right) =$

✎ **Condense each expression to a single logarithm.**

14) $\log 7 - \log 12 =$

15) $\log 8 + \log 3 =$

16) $4\log 2 - 7\log 5 =$

17) $6\log 4 - 9\log 5 =$

18) $3\log 8 - \log 17 =$

19) $8\log 3 - 6\log 2 =$

20) $\log 11 - 2\log 5 =$

21) $4\log 6 + 3\log 9 =$

22) $4\log 5 + 5\log 13 =$

23) $7\log_5 a + 16\log_5 b =$

24) $5\log_6 x - 7\log_6 y =$

25) $\log_5 u - 9\log_5 v =$

26) $8\log_3 u + 21\log_3 v =$

27) $26\log_7 u - 15\log_7 v =$

Natural Logarithms

✎ **Solve each equation for** x.

1) $e^x = 9$

2) $e^x = 36$

3) $e^x = 49$

4) $\ln x = 3$

5) $\ln(\ln x) = 3$

6) $e^x = 11$

7) $\ln(5x + 9) = 1$

8) $\ln(7x + 3) = 3$

9) $\ln(8x + 5) = 4$

10) $\ln x = \frac{1}{3}$

11) $\ln 6x = e^4$

12) $\ln x = \ln 4 + \ln 7$

13) $\ln x = 3\ln 3 + \ln 8$

✎ **Evaluate without using a calculator.**

14) $3\ln e =$

15) $\ln e^{10} =$

16) $4 \ln e =$

17) $\ln e^{15} =$

18) $13\ln e =$

19) $3\ln e^4 =$

20) $e^{\ln 19} =$

21) $e^{2\ln 5} =$

22) $e^{4\ln 3} =$

23) $\ln \sqrt[6]{e} =$

✎ **Reduce the following expressions to simplest form.**

24) $e^{-2\ln 6 + 2\ln 4} =$

25) $e^{-2\ln\left(\frac{4}{5e}\right)} =$

26) $2\ln(e^6) =$

27) $\ln\left(\frac{1}{e}\right)^9 =$

28) $e^{\ln 6 + 3\ln 5} =$

29) $e^{\ln\left(\frac{13}{e}\right)} =$

30) $7\ln(1^{-3e}) =$

31) $\ln\left(\frac{1}{e}\right)^{-12} =$

32) $3\ln\left(\frac{\sqrt[6]{e}}{3e}\right) =$

33) $e^{-3\ln e + 3\ln 3} =$

34) $e^{\ln\frac{15}{e}} =$

35) $19\ln(e^e) =$

Exponential Equations and Logarithms

✍ **Solve each equation for the unknown variable.**

1) $3^{2n} = 27$

2) $5^r = 125$

3) $15^n = 85$

4) $8^{r+3} = 2$

5) $144^x = 12$

6) $7^{-3v-2} = 49$

7) $8^{2n} = 64$

8) $6^n = 1,296$

9) $\dfrac{15^{2a}}{3^{-a}} = 315$

10) $11 \times 11^{-v} = 1,331$

11) $3^{2n} = \dfrac{1}{81}$

12) $\left(\dfrac{1}{9}\right)^n = 81$

13) $256^{2x} = 4$

14) $9^{3-2x} = 9^{-x}$

15) $6^{-3x} = 6^{x-3}$

16) $2^{3n} = 32$

17) $12^{5x+3} = 12^{2x}$

18) $10^{2n} = 100$

19) $3^{-4k} = 243$

20) $3^r = 9^{-4r}$

21) $13^{x+3} = 13^{4x}$

22) $9^{3x} = 729$

23) $15 \times 15^{-v} = 225$

24) $\dfrac{81}{3^{-2m}} = 3^{-2m-1}$

25) $8^{-2n} \times 8^2 = 8^{-n}$

26) $\left(\dfrac{1}{9}\right)^{2n+1} \times \left(\dfrac{1}{9}\right)^{-n-10} = \left(\dfrac{1}{9}\right)^{-2n}$

✍ **Solve each problem. (Round to the nearest whole number)**

27) A substance decays 15% each day. After 11 days, there are 6 milligrams of the substance remaining. How many milligrams were there initially? _____

28) A culture of bacteria grows continuously. The culture doubles every 4 hours. If the initial number of bacteria is 13, how many bacteria will there be in 23 hours? _____

29) Bob plans to invest $12,000 at an annual rate of 6.5%. How much will Bob have in the account after six years if the balance is compounded quarterly? _____

30) Suppose you plan to invest $8,000 at an annual rate of 6%. How much will you have in the account after 4 years if the balance is compounded monthly? _____

Solving Logarithmic Equations

✎ **Find the value of the variables in each equation.**

1) $\log(x) + 8 = 4$

2) $-\log_3 4x = 5$

3) $\log(x) + 7 = 6$

4) $\log x - \log 7 = 4$

5) $\log x + \log 4 = 2$

6) $\log 4 + \log x = 3$

7) $\log x + \log 2 = \log 12$

8) $-3\log_3 (x - 2) = -15$

9) $\log 4x = \log (3x + 2)$

10) $\log (2k - 4) = \log (k - 5)$

11) $\log(5p - 2) = \log(-2p + 12)$

12) $-8 + \log_3 (n + 3) = -8$

13) $\log_3 (x + 5) = \log_3 (x^2 + 8)$

14) $\log_9 (v^2 + 24) = \log_9 (-3v - 8)$

15) $\log (9 + 4b) = \log (7b^2 + 6b)$

16) $\log_9 (x + 8) - \log_9 x = \log_9 7$

17) $\log_5 9 + \log_5 x^2 = \log_5 81$

18) $\log_6 (x + 5) + \log_6 x = \log_6 24$

✎ **Find the value of x in each natural logarithm equation.**

19) $\ln 9 - \ln(3x + 9) = 3$

20) $\ln(x - 4) - \ln(x - 3) = \ln 4$

21) $\ln e^{27} - \ln(x + 3) = 3$

22) $\ln(2x - 6) - \ln(x - 12) = \ln 10$

23) $\ln 6x + \ln(x - 2) = \ln 3x$

24) $\ln(x - 3) - 2\ln(x - 3) = \ln 9$

25) $\ln (9x + 3) - \ln 5 = 6$

26) $\ln(x - 5) + \ln(x - 4) = \ln 2$

27) $\ln 8 + \ln(x + 4) = 10$

28)

29) $3 \ln 3x - \ln(x + 9) = 3 \ln 3x$

30) $\ln x^2 + \ln x^4 = \ln 1$

31) $\ln x^6 - \ln(x + 6) = 6 \ln x$

32) $16 \ln(x - 2) = 4 \ln(x^2 - 4x + 4)$

33) $\ln(x^2 + 10) = \ln(3x + 8)$

34) $6 \ln x - 6\ln(x + 3) = 12\ln(x^2)$

35) $\ln(2x - 3) - \ln(4x - 3) = \ln 4$

36) $\ln 3 + 9 \ln(x + 2) = \ln 3$

37) $3\ln e^2 + \ln(3x - 2) = \ln 3 + 9$

Answers of Worksheets – Chapter 12

Rewriting Logarithms

1) $5^2 = 25$
2) $4^4 = 256$
3) $3^4 = 81$
4) $8^2 = 64$
5) $6^3 = 216$
6) $2^4 = 16$
7) $10^2 = 100$
8) $3^5 = 243$
9) $5^4 = 625$
10) $2^8 = 256$
11) $3^8 = 6,561$
12) $11^2 = 121$
13) $14^2 = 196$
14) $81^{\frac{1}{4}} = 3$
15) $27^{\frac{1}{3}} = 3$
16) $32^{\frac{1}{5}} = 2$
17) $512^{\frac{1}{3}} = 8$
18) $2^{-3} = \frac{1}{8}$
19) $2^{-4} = \frac{1}{16}$
20) $a^b = \frac{7}{3}$
21) $\log_{12} 144 = 2$
22) $\log_3 343 = 7$
23) $\log_4 1,024 = 5$
24) $\log_{15} 225 = 2$
25) $\log_5 625 = 4$
26) $\log_6 1,296 = 4$
27) $\log_2 512 = 9$
28) $\log_5 3,125 = 5$
29) $\log_4 \frac{1}{4,096} = -6$
30) $\log_3 \frac{1}{243} = -5$
31) $\log_{16} \frac{1}{256} = -2$
32) $\log_6 \frac{1}{216} = -3$
33) $\log_3 \frac{1}{19,683} = -9$
34) $\log_{21} \frac{1}{441} = -2$

Evaluating Logarithms

1) 7
2) 8
3) 3
4) 4
5) 5
6) 5
7) 2
8) -1
9) -2
10) -4
11) -3
12) -4
13) -2
14) $-\frac{1}{4}$
15) $\frac{1}{3}$
16) -6
17) -3
18) $-\frac{1}{3}$
19) $(4, 6)$
20) $(\frac{1}{4}, -6)$
21) $(7, 2)$
22) $(6, 5)$
23) $(4, 2)$
24) $(7, -4)$
25) $(3, 3)$

Properties of Logarithms

1) $\log 9 + \log 4$
2) $\log 6 + \log 3$
3) $\log 2 + \log 8$
4) $\log 8 - \log 7$
5) $\log 9 - \log 5$
6) $3 \log 4 - 3 \log 11$
7) $\log 9 + 3 \log 4$
8) $2\log 7 - 2 \log 3$

ACT Math Workbook

9) $4\log 5 - \log 9$

10) $7\log x + 7\log y$

11) $2\log x + \log y + 5\log z$

12) $8\log u - \log v$

13) $\log x - 4\log y$

14) $\log \frac{7}{12}$

15) $\log(8 \times 3)$

16) $\log \frac{2^4}{5^7}$

17) $\log \frac{4^6}{9^5}$

18) $\log \frac{8^3}{17}$

19) $\log \frac{3^8}{2^6}$

20) $\log \frac{11}{5^2}$

21) $\log (6^4 \times 9^3)$

22) $\log (5^4 \times 13^5)$

23) $\log_5 (a^7 b^{16})$

24) $\log_6 \frac{x^5}{y^7}$

25) $\log_5 \frac{u}{v^9}$

26) $\log_3 (u^8 \times v^{21})$

27) $\log_7 \frac{u^{26}}{v^{15}}$

Natural Logarithms

1) $x = \ln 9$

2) $x = \ln 36, x = 2\ln(6)$

3) $x = \ln 49, x = 2\ln(7)$

4) $x = e^3$

5) $x = e^{e^3}$

6) $x = \ln 11$

7) $x = \frac{e-9}{5}$

8) $x = \frac{e^3-3}{7}$

9) $x = \frac{e^4-5}{8}$

10) $x = \sqrt[3]{e}$

11) $x = \frac{e^{e^4}}{6}$

12) $x = 28$

13) $x = 216$

14) 3

15) 10

16) 4

17) 15

18) 13

19) 12

20) 19

21) 25

22) 81

23) $\frac{1}{6}$

24) $\frac{4}{9}$

25) $\frac{25}{16e^2}$

26) 12

27) -9

28) 750

29) $\frac{13}{e}$

30) 0

31) 12

32) -5.8

33) $27e^{-3} = \frac{27}{e^3}$

34) $\frac{15}{e}$

35) $19e$

Exponential Equations and Logarithms

1) $\frac{3}{2}$

2) 3

3) 1.64

4) $\frac{-8}{3}$

5) $\frac{1}{2}$

6) $-\frac{4}{3}$

7) 1

8) 0.883

9) -2
10) -2
11) -2
12) $\frac{1}{8}$
13) 3
14) $\frac{3}{4}$
15) $\frac{5}{3}$

16) -1
17) 1
18) $-\frac{5}{4}$
19) 0
20) 1
21) 1
22) -1
23) -1.25

24) 2
25) 3
26) 35.9
27) 699.6
28) $\$17,668.3$
29) $\$10,163.9$

Solving Logarithmic Equations

1) $\{\frac{1}{10,000}\}$
2) $\{\frac{1}{972}\}$
3) $\{\frac{1}{10}\}$
4) $\{70,000\}$
5) $\{25\}$
6) $\{250\}$
7) $\{6\}$
8) $\{245\}$
9) $\{2\}$
10) No Solution
11) $\{2\}$
12) $\{-2\}$
13) No Solution

14) No Solution
15) $\{1, -\frac{9}{7}\}$
16) $\{\frac{4}{3}\}$
17) $\{3, -3\}$
18) $\{3\}$
19) $x = \frac{3-3e^3}{e^3}$
20) No Solution
21) $e^{24} - 3$
22) $\{\frac{57}{4}\}$
23) $\{\frac{5}{2}\}$
24) $\{\frac{28}{9}\}$

25) $x = \frac{5e^6-3}{9}$
26) $x = 6$
27) $x = \frac{e^{10}-32}{8}$
28) No Solution
29) $\{1, -1\}$
30) No Solution
31) $x = 3$
32) $\{1, 2\}$
33) $\{0.64951 \dots\}$
34) No Solution
35) $\{-1\}$
36) $x = \frac{3e^3+2}{3}$

Chapter 13:
Geometry and Solid Figures

Topics that you will practice in this chapter:

- ✓ Angles
- ✓ Pythagorean Relationship
- ✓ Triangles
- ✓ Polygons
- ✓ Trapezoids
- ✓ Circles
- ✓ Cubes
- ✓ Rectangular Prism
- ✓ Cylinder
- ✓ Pyramids and Cone

Mathematics is, as it were, a sensuous logic, and relates to philosophy as do the arts, music, and plastic art to poetry. — K. Shegel

Angles

✎ **What is the value of x in the following figures?**

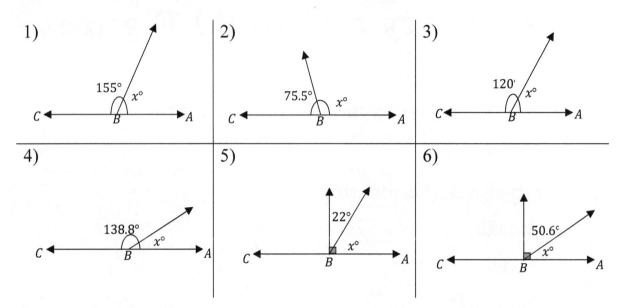

✎ **Calculate.**

7) Two supplementary angles have equal measures. What is the measure of each angle? _____

8) The measure of an angle is nine seventh the measure of its supplement. What is the measure of the angle? _____

9) Two angles are complementary and the measure of one angle is 24 less than the other. What is the measure of the bigger angle? _____

10) Two angles are complementary. The measure of one angle is one fifth the measure of the other. What is the measure of the smaller angle? _____

11) Two supplementary angles are given. The measure of one angle is 80° less than the measure of the other. What does the bigger angle measure? _____

Pythagorean Relationship

✎ **Do the following lengths form a right triangle?**

1)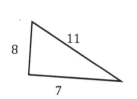

2)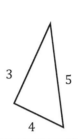

3)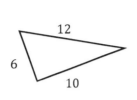

4)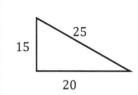

5)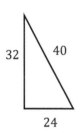

6)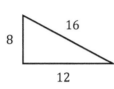

7)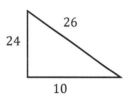

8)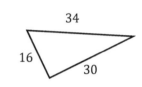

✎ **Find the missing side?**

9)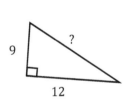

10)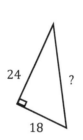

11)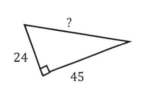

12)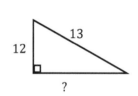

13)

14)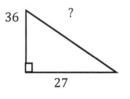

15)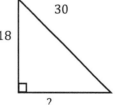

16)

Triangles

✎ Find the measure of the unknown angle in each triangle.

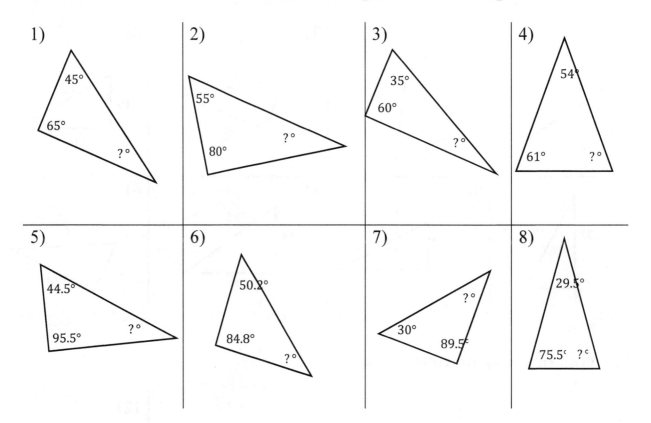

✎ Find area of each triangle.

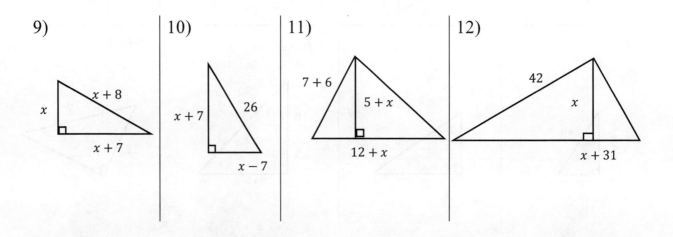

Polygons

✏️ **Find the perimeter of each shape.**

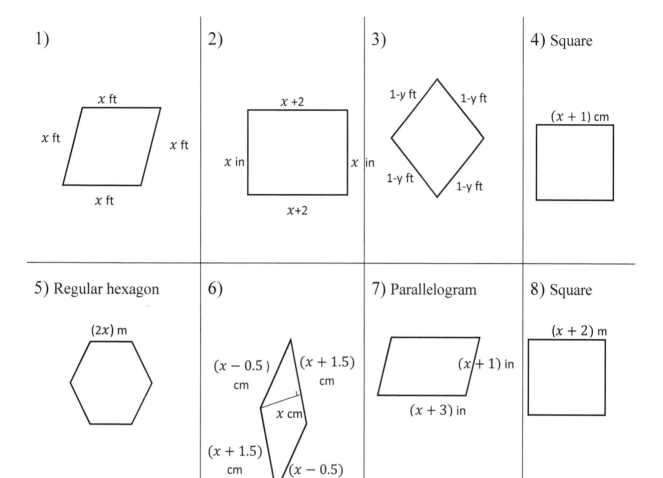

✏️ **Find the area of each shape.**

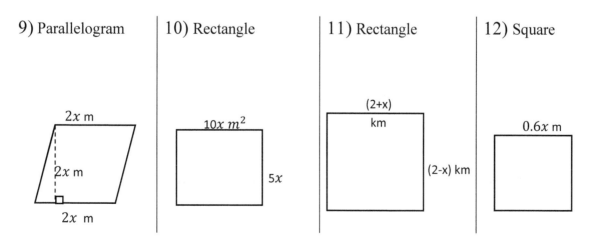

Trapezoids

Find the area of each trapezoid.

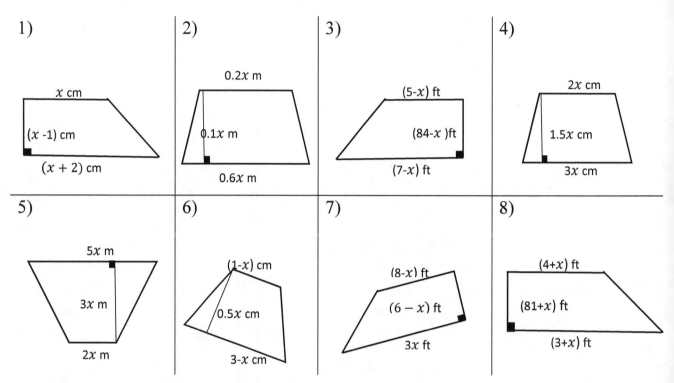

Calculate.

1) A trapezoid has an area of 40 cm² and its height is 8 cm and one base is 6 cm. What is the other base length? _____

2) If a trapezoid has an area of 85 ft² and the lengths of the bases are 9 ft and 8 ft, find the height. _____

3) If a trapezoid has an area of 150 m² and its height is 15 m and one base is 9 m, find the other base length. _____

4) The area of a trapezoid is 196 ft² and its height is 14 ft. If one base of the trapezoid is 12 ft, what is the other base length? _____

Circles

Find the area of each circle. ($\pi = 3.14$)

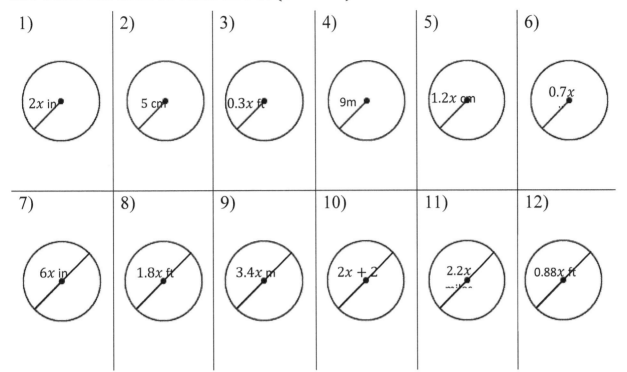

1) 2x in
2) 5 cm
3) 0.3x ft
4) 9m
5) 1.2x cm
6) 0.7x
7) 6x in
8) 1.8x ft
9) 3.4x m
10) 2x + 2
11) 2.2x miles
12) 0.88x ft

Complete the table below. ($\pi = 3.14$)

Circle No.	Radius	Diameter	Circumference	Area
1	1.4 inches	2.8 inches	8.792 inches	6.154 square inches
2		4.6 meters		
3				$2.01x^2$ square ft
4			36.42 miles	
5		6.2x kilometers		
6	5x centimeters			
7		2x feet		
8				1.54 square meters
9			5.7x inches	
10	(1-x) feet			

ACT Math Workbook

Cubes

✎ Find the volume of each cube.

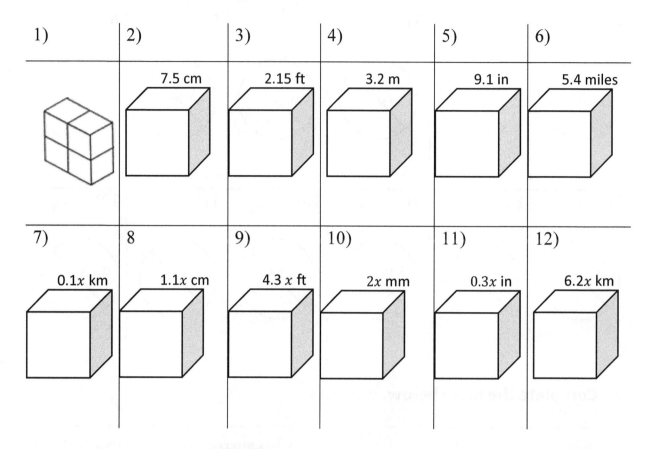

✎ Find the surface area of each cube.

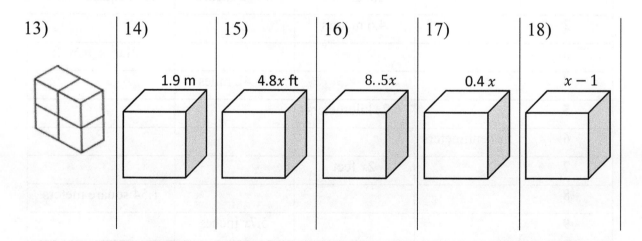

Rectangular Prism

✎ Find the volume of each Rectangular Prism.

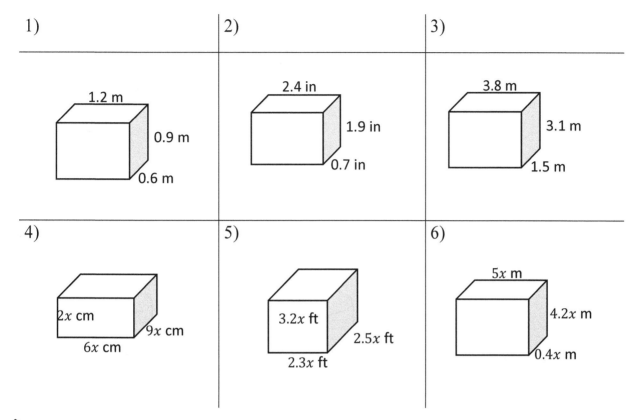

✎ Find the surface area of each Rectangular Prism.

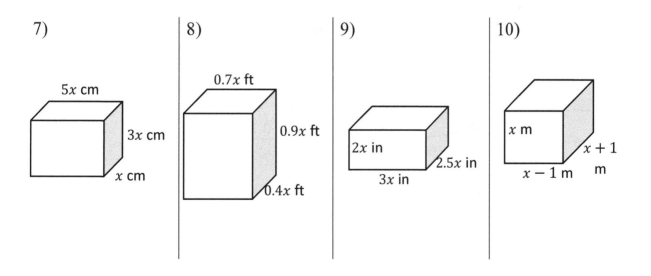

Cylinder

✎ **Find the volume of each Cylinder. Round your answer to the nearest tenth.** ($\pi = 3.14$)

1)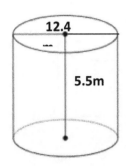

2)

3)

4)
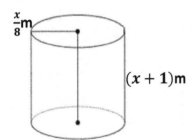

5)

6)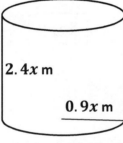

✎ **Find the surface area of each Cylinder.** ($\pi = 3.14$)

7)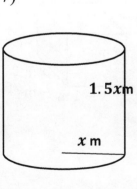

8)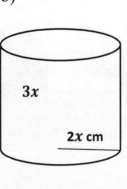

9)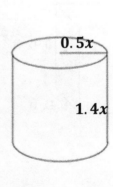

10)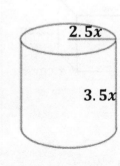

Pyramids and Cone

✎ **Find the volume of each Pyramid and Cone.** ($\pi = 3.14$)

1)
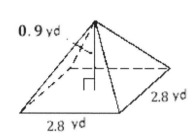

2)
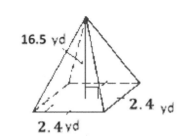

3)
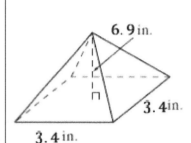

4)
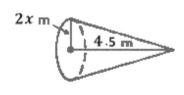

5)
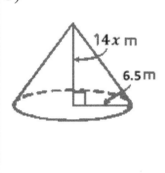

6)
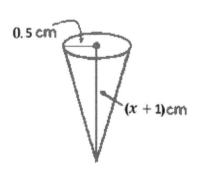

✎ **Find the surface area of each Pyramid and Cone.** ($\pi = 3.14$)

7)
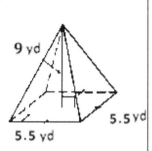

8)
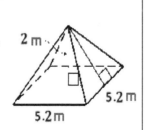

9)
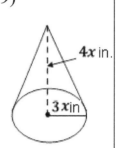

10)
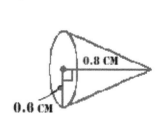

Answers of Worksheets – Chapter 13

Angles

1) 25°   4) 41.2°   7) 90°   10) 15°
2) 104.5°   5) 68°   8) 101.25°   11) 130°
3) 60°   6) 39.4°   9) 57°

Pythagorean Relationship

1) No   5) Yes   9) 15   13) 12
2) Yes   6) No   10) 30   14) 45
3) No   7) Yes   11) 51   15) 24
4) Yes   8) Yes   12) 5   16) 16

Triangles

1) 70°   6) 45°   10) $(\frac{x^2-49}{2})$ square unites
2) 45°   7) 60.5°   11) $(\frac{x^2+17x+60}{2})$ square unites
3) 85°   8) 75°   12) $(\frac{x^2+31x}{2})$ square unites
4) 65°   9) $(\frac{x^2+7x}{2})$ square unites
5) 40°

Polygons

1) $(4x)\ ft$ 5) $(12x)\ m$ 9) $(4x^2)\ m^2$
2) $(4x+4)\ in$ 6) $(4x+2)\ cm$ 10) $(50x^2)\ m^2$
3) $(4-4y)\ ft$ 7) $(4x+8)\ in$ 11) $(4-x^2)\ km^2$
4) $(4x+4)\ cm$ 8) $(4x+8)\ m$ 12) $(0.36x^2)\ m^2$

Trapezoids

1) $(x^2-1)\ cm^2$ 4) $(3.75x^2)\ cm^2$ 7) $(-x^2+2x+24)\ ft^2$
2) $(0.04x^2)\ m^2$ 5) $(10.5x^2)\ m^2$ 8) $(\frac{2x^2+9x+7}{2})\ ft^2$
3) $(x^2-10x+24)\ ft^2$ 6) $(x-0.5x^2)\ cm^2$

Calculate

1) 4 cm   2) 10 ft   3) 11 m   4) 16 ft

Circles

1) $(12.56x^2)\ in^2$ 4) $254.34\ m^2$ 7) $(28.56x^2)\ in^2$
2) $78.5\ cm^2$ 5) $(4.522x^2)\ cm^2$ 8) $(2.543x^2)\ ft^2$
3) $(0.283x^2)\ ft^2$ 6) $(1.54x^2)\ miles^2$ 9) $(9.075x^2)\ m^2$

ACT Math Workbook

10) $(3.14x^2 + 6.28x + 3.14)\ cm^2$ 11) $(3.8x^2)\ miles^2$ 12) $(0.608x^2)\ ft^2$

Circle No.	Radius	Diameter	Circumference	Area
1	1.4 inches	2.8 inches	8.792 inches	6.154 square inches
2	2.3 meters	4.6 meters	14.44 meters	16.61 meters
3	$0.8x$ square ft	$1.6x$ square ft	$5.024x$ square ft	$2.01x^2$ square ft
4	5.8 miles	11.6 miles	36.42 miles	105.63 miles
5	$3.1x$ kilometers	$6.2x$ kilometers	$19.47x$ kilometers	$30.175x^2$ kilometers
6	$5x$ centimeters	$10x$ centimeters	$31.4x$ centimeters	$78.5x^2$ centimeters
7	x feet	$2x$ feet	$6.28x$ feet	$3.14x^2$ feet
8	0.7 square meters	1.4 square meters	4.396 square meters	1.54 square meters
9	$2.5x$ inches	$5x$ inches	$15.7x$ inches	$19.625x^2$ inches
10	$(1-x)$ feet	$2 - 2x$ feet	$6.28 - 6.28x)$ feet	$3.14x^2 - 6.28x + 3.14)$ feet

Cubes

1) 4
2) $421.88\ cm^3$
3) $9.94\ ft^3$
4) $32.77\ m^3$
5) $753.57\ in^3$
6) $157.46\ miles^3$
7) $(0.001x^3)\ km^3$
8) $(1.33x^3)\ cm^3$
9) $(79.51x^3)\ ft^3$
10) $(8x^3)\ mm^3$
11) $(0.027x^3)\ in^3$
12) $(238.33x^3)\ km^3$
13) 12
14) $21.66\ m^2$
15) $(138.24x^2)\ ft^2$
16) $(433.5x^2)\ mm^2$
17) $(0.96x^2)\ km^2$
18) $6x^2 - 12x + 6\ cm^2$

Rectangular Prism

1) $0.65\ m^3$
2) $3.19\ in^3$
3) $17.67\ m^3$
4) $(108x^3)\ cm^3$
5) $(18.4x^3)\ ft^3$
6) $(8.4x^3)\ m^3$
7) $(46x^2)\ cm^2$
8) $(2.54x^2)\ ft^2$
9) $(37x^2)\ in^2$
10) $(6x^2 - 2)\ m^2$

Cylinder

1) $663.86\ m^3$
2) $632.6\ cm^3$
3) $(8,488.55x^2)\ cm^3$
4) $(0.05x^3 + 0.05x^2)\ m^3$
5) $(6.104x^3)\ m^3$
6) $(0.5\ x^3 + 1.51x^2)in^3$
7) $(15.7x^2)\ m^2$
8) $(62.8x^2)\ cm^2$
9) $(5.97x^2)\ cm^2$
10) $(94.2x^2)\ m^2$

Pyramids and Cone

1) $2.35\ yd^3$
2) $31.68\ yd^3$
3) $26.59\ in^3$
4) $(18.84x^2)\ m^3$
5) $(619.10x)\ m^3$
6) $(0.262x + 0.262)\ cm^3$
7) $133.77 yd^2$
8) $61.15\ m^2$
9) $(75.36x^2)in^2$
10) $(3.01)cm^2$

Chapter 14:
Trigonometric Functions

Topics that you will practice in this chapter:

- ✓ Trig ratios of General Angles
- ✓ Sketch Each Angle in Standard Position
- ✓ Finding Co-Terminal Angles and Reference Angles
- ✓ Angles in Radians
- ✓ Angles in Degrees
- ✓ Evaluating Each Trigonometric Expression
- ✓ Missing Sides and Angles of a Right Triangle
- ✓ Arc Length and Sector Area

Mathematics is like checkers in being suitable for the young, not too difficult, amusing, and without peril to the state. — Plato

Trig Ratios of General Angles

✎ **Evaluate.**

1) $\sin -135° =$ _____

2) $\sin 300° =$ _____

3) $\cos -390° =$ _____

4) $\cos 240° =$ _____

5) $\sin 390° =$ _____

6) $\sin -330° =$ _____

7) $\tan 120° =$ _____

8) $\cot 150° =$ _____

9) $\tan 210° =$ _____

10) $\cot 225° =$ _____

11) $\sec 330° =$ _____

12) $\csc 450° =$ _____

13) $\cot -135° =$ _____

14) $\sec 360° =$ _____

15) $\cos -450° =$ _____

16) $\sec 150° =$ _____

17) $\csc 360° =$ _____

18) $\cot -120° =$ _____

✎ **Find the exact value of each trigonometric function. Some may be undefined.**

19) $\sec 5\pi =$ _____

20) $\tan -\dfrac{5\pi}{2} =$ _____

21) $\cos \dfrac{4\pi}{2} =$ _____

22) $\cot \dfrac{20\pi}{6} =$ _____

23) $\sec -\dfrac{21\pi}{6} =$ _____

24) $\sec \dfrac{4\pi}{3} =$ _____

25) $\csc \dfrac{17\pi}{3} =$ _____

26) $\cot \dfrac{3\pi}{4} =$ _____

27) $\csc -\dfrac{7\pi}{6} =$ _____

28) $\cot \dfrac{2\pi}{3} =$ _____

Sketch Each Angle in Standard Position

✎ **Draw each angle with the given measure in standard position.**

1) 390°

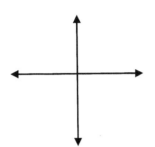

2) 750°

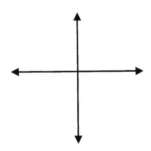

3) −300°

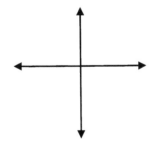

4) 1,100°

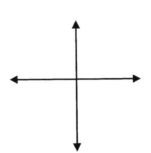

5) $-\frac{11\pi}{6}$

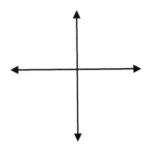

6) $\frac{23\pi}{6}$

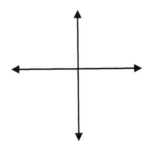

Finding Co-terminal Angles and Reference Angles

✎ **Find a conterminal angle between 0° and 360° for each angle provided.**

1) $-150° =$

2) $-270° =$

3) $-315° =$

4) $-810° =$

✎ **Find a conterminal angle between 0 and 2π for each given angle.**

5) $\dfrac{12\pi}{5} =$

6) $-\dfrac{19\pi}{8} =$

7) $-\dfrac{25\pi}{11} =$

8) $\dfrac{21\pi}{4} =$

✎ **Find the reference angle of each angle.**

9)
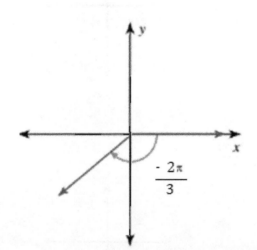
$-\dfrac{2\pi}{3}$

10)
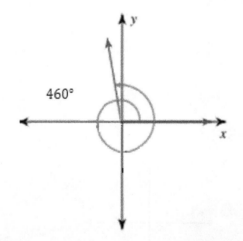
$460°$

Angles and Angle Measure

✎ **Convert each degree measure into radians.**

1) $225° =$ ___

2) $432° =$ ___

3) $840° =$ ___

4) $576° =$ ___

5) $420° =$ ___

6) $810° =$ ___

7) $-270° =$ ___

8) $855° =$ ___

9) $330° =$ ___

10) $288° =$ ___

11) $100° =$ ___

12) $420° =$ ___

13) $-150° =$ ___

14) $-280° =$ ___

15) $-260° =$ ___

16) $630° =$ ___

17) $-960° =$ ___

18) $864° =$ ___

19) $-144° =$ ___

20) $1,170° =$ ___

21) $810° =$ ___

✎ **Convert each radian measure into degrees.**

22) $\frac{\pi}{10} =$

23) $\frac{2\pi}{18} =$

24) $\frac{8\pi}{9} =$

25) $\frac{6\pi}{15} =$

26) $-\frac{17\pi}{10} =$

27) $\frac{5\pi}{18} =$

28) $-\frac{8\pi}{15} =$

29) $\frac{15\pi}{18} =$

30) $\frac{7\pi}{60} =$

31) $\frac{18\pi}{20} =$

32) $-\frac{3\pi}{18} =$

33) $\frac{13\pi}{90} =$

34) $-\frac{\pi}{9} =$

35) $\frac{12\pi}{30} =$

36) $-\frac{8\pi}{72} =$

37) $\frac{7\pi}{6} =$

38) $-\frac{8\pi}{40} =$

39) $-\frac{11\pi}{90} =$

Evaluating Trigonometric Functions

✎ **Find the exact value of each trigonometric function.**

1) $\cos 510° =$ _____

2) $\tan \frac{5\pi}{3} =$ _____

3) $\tan -\frac{11\pi}{2} =$ _____

4) $\cot -\frac{7\pi}{3} =$ _____

5) $\cos -\frac{19\pi}{4} =$ _____

6) $\cos -315° =$ _____

7) $\sin 405° =$ _____

8) $\tan 450° =$ _____

9) $\cot -480° =$ _____

10) $\tan 495° =$ _____

11) $\cot 810° =$ _____

12) $\sin -540° =$ _____

13) $\cot -585° =$ _____

✎ **Use the given point on the terminal side of angle θ to find the value of the trigonometric function indicated.**

14) $\sin\theta$; $(-6, 8)$

15) $\cos\theta$; $(-6, 8)$

16) $\cot\theta$; $(-2, -6)$

17) $\cos\theta$; $(10, 24)$

18) $\sin\theta$; $(9, -9)$

19) $\tan\theta$; $(-6, -\sqrt{12})$

Missing Sides and Angles of a Right Triangle

✎ Find the value of each trigonometric ratio as fractions in their simplest form.

1) tan x

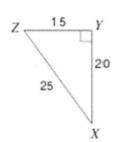

2) sin A

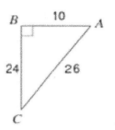

✎ Find the missing sides. Round answers to the nearest tenth.

3)

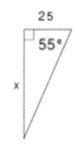

4)

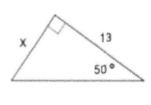

5)

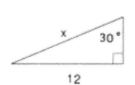

6)

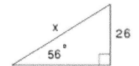

Arc Length and Sector Area

✏️ **Find the length of each arc. Round your answers to the nearest tenth.**

($\pi = 3.14$)

1) $r = 40$ cm, $\theta = 90°$

2) $r = 16$ ft, $\theta = 55°$

3) $r = 24$ ft, $\theta = 100°$

4) $r = 18$ m, $\theta = 85°$

✏️ **Find area of each sector. Do *not* round. Round your answers to the nearest tenth.** ($\pi = 3.14$)

5)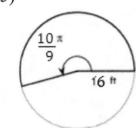

7)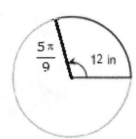

6)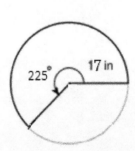

8)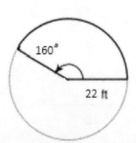

Answers of Worksheets – Chapter 14

Trig Ratios of General Angles

1) $-\frac{\sqrt{2}}{2}$
2) $-\frac{\sqrt{3}}{2}$
3) $\frac{\sqrt{3}}{2}$
4) $-\frac{1}{2}$
5) $\frac{1}{2}$
6) $\frac{1}{2}$
7) $-\sqrt{3}$
8) $-\sqrt{3}$
9) $\frac{\sqrt{3}}{3}$
10) 1
11) $\frac{2\sqrt{3}}{3}$
12) 1
13) 1
14) 1
15) 0
16) $-\frac{2\sqrt{3}}{3}$
17) Undefined
18) $\frac{\sqrt{3}}{3}$
19) -1
20) Undefined
21) 1
22) $\frac{\sqrt{3}}{3}$
23) Undefined
24) -2
25) $-\frac{2\sqrt{3}}{3}$
26) -1
27) 2
28) $-\frac{\sqrt{3}}{3}$

Sketch Each Angle in Standard Position

1) 390°
2) 750°
3) −300°

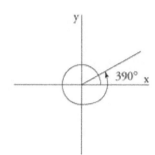
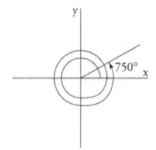
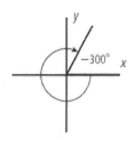

4) 1,110°
5) $-\frac{11\pi}{6} = -330°$
6) $\frac{23\pi}{6} = -690°$

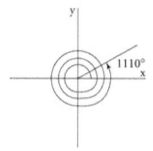
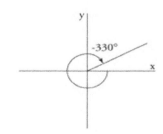
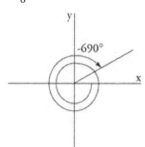

Finding Co-Terminal Angles and Reference Angles

1) 210°
2) 90°
3) 45°
4) 270°
5) $\frac{2\pi}{5}$
6) $\frac{13\pi}{8}$
7) $\frac{19\pi}{11}$
8) $\frac{5\pi}{4}$
9) $\frac{4\pi}{3}$
10) 100°

ACT Math Workbook

Angles and Angle Measure

1) $\frac{5\pi}{4}$
2) $\frac{12\pi}{5}$
3) $\frac{14\pi}{3}$
4) $\frac{16\pi}{5}$
5) $\frac{7\pi}{3}$
6) $\frac{9\pi}{2}$
7) $-\frac{3\pi}{2}$
8) $\frac{19\pi}{4}$
9) $\frac{11\pi}{6}$
10) $\frac{8\pi}{5}$
11) $\frac{5\pi}{9}$
12) $\frac{21\pi}{9}$
13) $-\frac{5}{6}\pi$
14) $-\frac{14\pi}{9}$
15) $-\frac{13\pi}{9}$
16) $\frac{7\pi}{2}$
17) $-\frac{16\pi}{3}$
18) $\frac{24\pi}{5}$
19) $-\frac{4\pi}{5}$
20) $\frac{13\pi}{2}$
21) $\frac{27\pi}{6}$
22) 18°
23) 20°
24) 160°
25) 72°
26) −306°
27) 50°
28) −96°
29) 150°
30) 21°
31) 162°
32) −30°
33) 26°
34) −20°
35) 72°
36) −20°
37) 210°
38) −36°
39) −22°

Evaluating Each Trigonometric Functions

1) $-\frac{\sqrt{3}}{2}$
2) $-\sqrt{3}$
3) Undefined
4) $-\frac{\sqrt{3}}{3}$
5) $-\frac{\sqrt{2}}{2}$
6) $\frac{\sqrt{2}}{2}$
7) $\frac{\sqrt{2}}{2}$
8) Undefined
9) $\frac{\sqrt{3}}{3}$
10) −1
11) 0
12) 0
13) −1
14) 0.6
15) −0.8
16) 3
17) $\frac{10}{26}$
18) $-\frac{\sqrt{2}}{2}$
19) $\sqrt{3}$

Missing Sides and Angles of a Right Triangle

1) $\frac{3}{4}$
2) $\frac{12}{13}$
3) 35.7
4) 10
5) 24
6) 31.4

Arc Length and Sector Area

1) 62.8 cm
2) 15.4 ft
3) 41.9 ft
4) 26.7 m
5) 446.6 ft^2
6) 567.2 in^2
7) 125.6 in^2
8) 675.4 ft^2

Chapter 15:
Statistics and Probability

Topics that you will practice in this chapter:

- ✓ Mean and Median
- ✓ Mode and Range
- ✓ Histograms
- ✓ Stem–and–Leaf Plot
- ✓ Pie Graph
- ✓ Probability Problems
- ✓ Factorials
- ✓ Combinations and Permutation

Mathematics is no more computation than typing is literature.

– John Allen Paulos

Mean and Median

✎ Find Mean and Median of the Given Data.

1) 8, 9, 19, 3, 4

2) 11, 7, 35, 10, 17, 32, 24

3) 38, 9, 15, 17, 13

4) 50, 19, 2, 18, 6, 7

5) 25, 27, 13, 16, 6, 13, 54

6) 24, 364, 42, 57, 6, 68

7) 89, 98, 65, 45, 3, 4, 30, 42

8) 34, 15, 15, 17, 22, 29, 15

9) 2, 5, 10, 45, 8, 13, 35, 6

10) 20, 22, 18, 7, 2, 17, 44, 53

11) 33, 52, 81, 9, 45, 31

12) 19, 74, 51, 8, 12, 15, 9, 14

✎ Calculate.

13) In a javelin throw competition, five athletics score 45, 33, 53, 46 and 19 meters. What are their Mean and Median? _____

14) Eva went to shop and bought 5 apples, 9 peaches, 4 bananas, 7 pineapples and 8 melons. What are the Mean and Median of her purchase? _____

15) Bob has 19 black pen, 15 red pen, 27 green pens, 21 blue pens and one boxes of yellow pens. If the Mean and Median are 19 respectively, what is the number of yellow pens in box? _____

Mode and Range

✎ Find Mode and Rage of the Given Data.

1) 7, 4, 18, 9, 9, 3
   Mode: _____   Range: _____

2) 8, 8, 15, 14, 8, 5, 6, 18
   Mode: _____   Range: _____

3) 4, 4, 4, 15, 19, 24, 31, 5, 4
   Mode: _____   Range: _____

4) 10, 10, 9, 17, 14, 8, 20, 4
   Mode: _____   Range: _____

5) 5, 11, 3, 4, 3, 3
   Mode: _____   Range: _____

6) 13, 7, 7, 7, 7, 4, 12, 25, 8, 3
   Mode: _____   Range: _____

7) 1, 7, 9, 9, 24, 24, 24, 20, 34, 35
   Mode: _____   Range: _____

8) 9, 4, 7, 13, 13, 13, 9, 8, 15
   Mode: _____   Range: _____

9) 8, 8, 8, 5, 8, 7, 17, 16, 3, 9
   Mode: _____   Range: _____

10) 34, 34, 32, 14, 6, 14, 9, 14
    Mode: _____   Range: _____

11) 8, 8, 6, 8, 18, 10, 16, 15
    Mode: _____   Range: _____

12) 12, 12, 7, 11, 14, 12, 33, 5
    Mode: _____   Range: _____

✎ Calculate.

13) A stationery sold 21 pencils, 42 red pens, 25 blue pens, 26 notebooks, 21 erasers, 28 rulers and 27 color pencils. What are the Mode and Range for the stationery sells?

    Mode: _____   Range: _____

14) In an English test, eight students score 19, 10, 10, 17, 35, 35, 14 and 10. What are their Mode and Range? _____

15) What is the range of the first 6 odd numbers greater than 8?
    _____

Times Series

✍ **Use the following Graph to complete the table.**

Day	Distance (km)
1	
2	

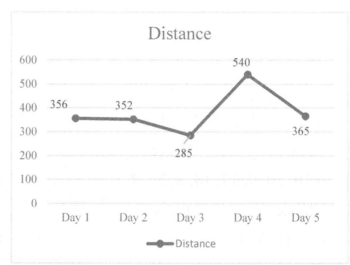

The following table shows the number of births in the US from 2007 to 2012 (in millions).

Year	Number of births (in millions)
2007	4.25
2008	4.19
2009	4.55
2010	3.80
2011	3.25
2012	2.54

Draw a Time Series for the table.

Stem-and-Leaf Plot

✎ **Make stem ad leaf plots for the given data.**

1) 41, 44, 47, 40, 70, 45, 79, 77, 49, 44, 19, 10

2) 21, 87, 56, 20, 27, 23, 55, 82, 82, 53, 87, 58

3) 111, 47, 66, 44, 94, 117, 62, 114, 48, 112, 68, 99

4) 52, 25, 101, 58, 71, 26, 109, 53, 75, 29, 53, 108, 79

5) 51, 88, 9, 87, 81, 8, 3, 50, 85, 54, 9, 54, 5

6) 40, 93, 20, 25, 48, 92, 95, 52, 21, 44, 97, 29

Pie Graph

The circle graph below shows all Robert's expenses for last month. Robert spent $384 on his hobbies last month.

Answer following questions based on the Pie graph.

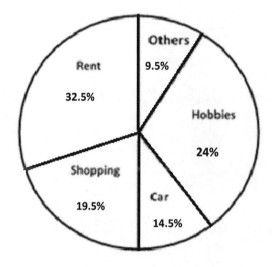

1) How much was Robert's total expenses last month? _____

2) How much did Robert spend on his car last month? _____

3) How much did Robert spend for shopping last month? _____

4) How much did Robert spend on his rent last month? _____

5) What fraction is Robert's expenses for his car and shopping out of his total expenses last month? _____

Probability Problems

✍ **Calculate.**

1) A number is chosen at random from 1 to 20. Find the probability of selecting number 8 or smaller numbers. _____

2) Bag A contains 16 red marbles and 6 green marbles. Bag B contains 12 black marbles and 18 orange marbles. What is the probability of selecting a green marble at random from bag A? What is the probability of selecting a black marble at random from Bag B? _____

3) A number is chosen at random from 1 to 25. What is the probability of selecting multiples of 5? _____

4) A card is chosen from a well-shuffled deck of 52 cards. What is the probability that the card will be a queen? _____

5) A number is chosen at random from 1 to 15. What is the probability of selecting a multiple of 4? _____

A spinner, numbered 1–8, is spun once. What is the probability of spinning …?

6) an Odd number? _____   7) a multiple of 2? _____

8) a multiple of 5? _____   9) number 10? _____

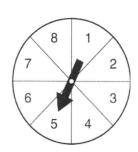

Factorials

✎ **Determine the value for each expression.**

1) $6! + 1! =$

2) $5! + 2! =$

3) $(4!)^2 =$

4) $6! - 3! =$

5) $8! - 4! + 3 =$

6) $3! \times 4 - 12 =$

7) $(3! + 1!)^2 =$

8) $(5! - 4!)^2 =$

9) $(3! \, 0!)^2 - 2 =$

10) $\dfrac{8!}{6!} =$

11) $\dfrac{3!}{2!} =$

12) $\dfrac{6!}{5!} =$

13) $\dfrac{21!}{19!} =$

14) $\dfrac{(n-1)!}{(n-3)!} =$

15) $\dfrac{(n+2)!}{(n+1)!} =$

16) $\dfrac{(4+2!)^3}{2!} =$

17) $\dfrac{4n!}{2n!} =$

18) $\dfrac{31!}{29!2!} =$

19) $\dfrac{13!}{9!3!} =$

20) $\dfrac{6 \times 280!}{3(4 \times 70)!} =$

21) $\dfrac{30!}{31!2!} =$

22) $\dfrac{7!7!}{8!5!} =$

23) $\dfrac{12!11!}{9!10!} =$

24) $\dfrac{(2 \times 5)!}{1!9!} =$

25) $\dfrac{2!(6n-1)!}{(6n)!} =$

26) $\dfrac{n(4n+4)!}{(4n+5)!} =$

27) $\dfrac{(n+1)!(n)}{(n+2)!} =$

ACT Math Workbook

Combinations and Permutations

✎ Calculate the value of each.

1) $6! = $ ___

2) $2! \times 5! = $ ___

3) $4! = $ ___

4) $3! + 5! = $ ___

5) $7! = $ ___

6) $9! = $ ___

7) $3! + 3! = $ ___

8) $5! - 2! = $ ___

✎ Find the answer for each word problems.

9) Susan is baking cookies. She uses sugar, Vanilla and eggs. How many different orders of ingredients can she try? ___

10) Albert is planning for his vacation. He wants to go to museum, watch a movie, go to the beach, play volleyball and play football. How many ways of ordering are there for him? ___

11) How many 6-digit numbers can be named using the digits 1, 6, 8, 9, and 10 without repetition? ___

12) In how many ways can 4 boys be arranged in a straight line? ___

13) In how many ways can 8 athletes be arranged in a straight line? ___

14) A professor is going to arrange her 5 students in a straight line. In how many ways can she do this? ___

15) How many code symbols can be formed with the letters for the word FRIEND? ___

16) In how many ways a team of 7 basketball players can to choose a captain and co-captain? ___

WWW.MathNotion.Com

Answers of Worksheets – Chapter 15

Mean and Median

1) Mean: 8.6, Median: 8
2) Mean: 19.43, Median: 17
3) Mean: 18.4, Median: 15
4) Mean: 17, Median: 12.5
5) Mean: 22, Median: 16
6) Mean: 93.5, Median: 49.5
7) Mean: 47, Median: 43.5
8) Mean: 21, Median: 17
9) Mean: 15.5, Median: 9
10) Mean: 22.88, Median: 19
11) Mean: 41.83, Median: 39
12) Mean: 25.25, Median: 14.5
13) Mean: 39.2, Median: 45
14) Mean: 6.6, Median: 7
15) 13

Mode and Range

1) Mode: 9, Range: 15
2) Mode: 8, Range: 13
3) Mode: 4, Range: 27
4) Mode: 10, Range: 16
5) Mode: 3, Range: 8
6) Mode: 7, Range: 22
7) Mode: 24, Range: 34
8) Mode: 13, Range: 11
9) Mode: 8, Range: 14
10) Mode: 14, Range: 28
11) Mode: 8, Range: 12
12) Mode: 12, Range: 28
13) Mode: 21, Range: 21
14) Mode: 10, Range: 25
15) 10

Time series

Day	Distance (km)
1	356
2	352
3	285
4	540
5	365

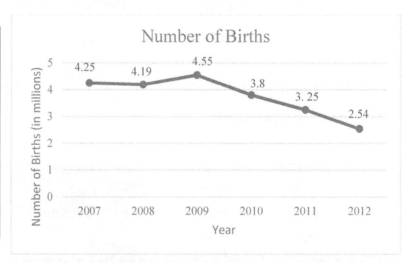

Stem–And–Leaf Plot

1)

Stem	leaf
1	0 9
4	0 1 4 4 5 7 9
7	0 7 9

2)

Stem	leaf
2	0 1 3 7
5	3 5 6 8
8	2 2 7 7

3)

Stem	leaf
4	4 7 8
6	2 6 8
9	4 9
11	1 2 4 7

4)

Stem	leaf
2	5 6 9
5	2 3 3 8
7	1 5 9
10	1 8 9

5)

Stem	leaf
0	3 5 8 9 9
5	0 1 4 4
8	1 5 7 8

6)

Stem	leaf
2	0 1 5 9
4	0 2 4 8
9	2 3 5 7

Pie Graph

1) $1,600
2) $232
3) $312
4) $520
5) $\frac{17}{50}$

Probability Problems

1) $\frac{2}{5}$
2) $\frac{3}{11}, \frac{2}{5}$
3) $\frac{1}{5}$
4) $\frac{1}{13}$
5) $\frac{1}{5}$
6) $\frac{1}{2}$
7) $\frac{1}{2}$
8) $\frac{1}{8}$
9) 0

Factorials

1) 721
2) 122
3) 576
4) 714
5) 40,299
6) 12
7) 49
8) 9,216
9) 34
10) 56
11) 3
12) 6
13) 420
14) $(n-1)(n-2)$
15) $n+2$
16) 108
17) 2
18) 465
19) 2,860
20) 2
21) $\frac{1}{62}$
22) 5.25
23) 14,520
24) 10
25) $\frac{1}{3n}$
26) $\frac{n}{4n+5}$
27) $\frac{n}{n+2}$

Combinations and Permutations

1) 720
2) 240
3) 24
4) 126
5) 5,040
6) 362,880
7) 12
8) 118
9) 6
10) 120
11) 720
12) 24
13) 40,320
14) 120
15) 720
16) 42

ACT Math Test Review

Since 1959, the American College Testing Organization (ACT) has been judging student's potential regarding academics. ACT is a standardized test used for college admissions in the United States. In essence, it is a broad and quick assessment of students' academic abilities. The ACT is divided into four major segments.

- English
- Reading
- Mathematics
- Science

The ACT also includes an optional 40-minute Writing Test.

In an ACT assessment test, all questions are weighted the same. You also have to keep in mind that the more difficult questions are randomly thrown around in the test. You can choose to skip over the more challenging tasks and ace out the simpler questions in the tests first.

There are 60 Mathematics questions on ACT and students have 60 minutes to answer the questions. The Mathematics section of the ACT contains multiple choice questions.

ACT Mathematics cover the following topics:

- Pre-Algebra (20-25%)
- Elementary Algebra (15-20%)
- Intermediate Algebra (15-20%)
- Coordinate Geometry (15-20%)
- Plane Geometry (20-25%)
- Trigonometry (5-10%)

ACT permits the use of personal calculators on the Math portion of the test.

In this section, there are two complete ACT Mathematics Tests. Take these tests to see what score you'll be able to receive on a real ACT test.

Time to Test

Time to refine your skill with a practice examination

Take a practice ACT Math Test to simulate the test day experience. After you've finished, score your test using the answer key.

Before You Start

- You'll need a pencil, a calculator and a timer to take the test.
- For each question, there are five possible answers. Choose which one is best.
- After you've finished the test, review the answer key to see where you went wrong.

Good Luck!

The hardest arithmetic to master is that which enables us to count our blessings.
~Eric Hoffer

ACT Math Practice Test Answer Sheets

Remove (or photocopy) these answer sheets and use them to complete the practice tests.

ACT Practice Test

1	Ⓐ Ⓑ Ⓒ Ⓓ Ⓔ	21	Ⓐ Ⓑ Ⓒ Ⓓ Ⓔ	41	Ⓐ Ⓑ Ⓒ Ⓓ Ⓔ
2	Ⓐ Ⓑ Ⓒ Ⓓ Ⓔ	22	Ⓐ Ⓑ Ⓒ Ⓓ Ⓔ	42	Ⓐ Ⓑ Ⓒ Ⓓ Ⓔ
3	Ⓐ Ⓑ Ⓒ Ⓓ Ⓔ	23	Ⓐ Ⓑ Ⓒ Ⓓ Ⓔ	43	Ⓐ Ⓑ Ⓒ Ⓓ Ⓔ
4	Ⓐ Ⓑ Ⓒ Ⓓ Ⓔ	24	Ⓐ Ⓑ Ⓒ Ⓓ Ⓔ	44	Ⓐ Ⓑ Ⓒ Ⓓ Ⓔ
5	Ⓐ Ⓑ Ⓒ Ⓓ Ⓔ	25	Ⓐ Ⓑ Ⓒ Ⓓ Ⓔ	45	Ⓐ Ⓑ Ⓒ Ⓓ Ⓔ
6	Ⓐ Ⓑ Ⓒ Ⓓ Ⓔ	26	Ⓐ Ⓑ Ⓒ Ⓓ Ⓔ	46	Ⓐ Ⓑ Ⓒ Ⓓ Ⓔ
7	Ⓐ Ⓑ Ⓒ Ⓓ Ⓔ	27	Ⓐ Ⓑ Ⓒ Ⓓ Ⓔ	47	Ⓐ Ⓑ Ⓒ Ⓓ Ⓔ
8	Ⓐ Ⓑ Ⓒ Ⓓ Ⓔ	28	Ⓐ Ⓑ Ⓒ Ⓓ Ⓔ	48	Ⓐ Ⓑ Ⓒ Ⓓ Ⓔ
9	Ⓐ Ⓑ Ⓒ Ⓓ Ⓔ	29	Ⓐ Ⓑ Ⓒ Ⓓ Ⓔ	49	Ⓐ Ⓑ Ⓒ Ⓓ Ⓔ
10	Ⓐ Ⓑ Ⓒ Ⓓ Ⓔ	30	Ⓐ Ⓑ Ⓒ Ⓓ Ⓔ	50	Ⓐ Ⓑ Ⓒ Ⓓ Ⓔ
11	Ⓐ Ⓑ Ⓒ Ⓓ Ⓔ	31	Ⓐ Ⓑ Ⓒ Ⓓ Ⓔ	51	Ⓐ Ⓑ Ⓒ Ⓓ Ⓔ
12	Ⓐ Ⓑ Ⓒ Ⓓ Ⓔ	32	Ⓐ Ⓑ Ⓒ Ⓓ Ⓔ	52	Ⓐ Ⓑ Ⓒ Ⓓ Ⓔ
13	Ⓐ Ⓑ Ⓒ Ⓓ Ⓔ	33	Ⓐ Ⓑ Ⓒ Ⓓ Ⓔ	53	Ⓐ Ⓑ Ⓒ Ⓓ Ⓔ
14	Ⓐ Ⓑ Ⓒ Ⓓ Ⓔ	34	Ⓐ Ⓑ Ⓒ Ⓓ Ⓔ	54	Ⓐ Ⓑ Ⓒ Ⓓ Ⓔ
15	Ⓐ Ⓑ Ⓒ Ⓓ Ⓔ	35	Ⓐ Ⓑ Ⓒ Ⓓ Ⓔ	55	Ⓐ Ⓑ Ⓒ Ⓓ Ⓔ
16	Ⓐ Ⓑ Ⓒ Ⓓ Ⓔ	36	Ⓐ Ⓑ Ⓒ Ⓓ Ⓔ	56	Ⓐ Ⓑ Ⓒ Ⓓ Ⓔ
17	Ⓐ Ⓑ Ⓒ Ⓓ Ⓔ	37	Ⓐ Ⓑ Ⓒ Ⓓ Ⓔ	57	Ⓐ Ⓑ Ⓒ Ⓓ Ⓔ
18	Ⓐ Ⓑ Ⓒ Ⓓ Ⓔ	38	Ⓐ Ⓑ Ⓒ Ⓓ Ⓔ	58	Ⓐ Ⓑ Ⓒ Ⓓ Ⓔ
19	Ⓐ Ⓑ Ⓒ Ⓓ Ⓔ	39	Ⓐ Ⓑ Ⓒ Ⓓ Ⓔ	59	Ⓐ Ⓑ Ⓒ Ⓓ Ⓔ
20	Ⓐ Ⓑ Ⓒ Ⓓ Ⓔ	40	Ⓐ Ⓑ Ⓒ Ⓓ Ⓔ	60	Ⓐ Ⓑ Ⓒ Ⓓ Ⓔ

ACT Practice Test 1

Mathematics

❖ **60 Questions.**

❖ **Total time for this test: 60 Minutes.**

❖ **You may use a scientific calculator on this test.**

Administered *Month Year*

ACT Math Workbook

1) $5^{\frac{5}{2}} \times 5^{\frac{1}{2}} = ?$

   A. 5^5 D. 5^6

   B. 5^4 E. 5^0

   C. 5^3

2) If $\frac{5x}{36} = \frac{x-2}{6}$, $x = ?$

   A. $\frac{1}{12}$ D. 12

                                        E. $\frac{12}{5}$

   B. $\frac{3}{5}$

   C. 6

3) 121 is equal to?

   A. $25 - (5 \times 8) + (7 \times 20)$

   B. $\left(\frac{10}{7} \times 63\right) + \left(\frac{124}{4}\right)$

   C. $\left(\left(\frac{70}{6} + \frac{22}{3}\right) \times 7\right) - \frac{20}{3} + \frac{130}{6}$

   D. $(3 \times 11) + (42 \times 2.5) - 14$

   E. $\frac{148}{8} + \frac{207}{2}$

4) Five years ago, Amy was three times as old as Mike was. If Mike is 11 years old now, how old is Amy?

   A. 33   D. 21

   B. 25   E. 23

   C. 16

5) A number is chosen at random from 1 to 15. Find the probability of not selecting a composite number.

A. $\frac{3}{15}$

B. 15

C. $\frac{2}{5}$

D. 2

E. 1

6) If $|a| < 3$ then which of the following is true? $(b > 0)$?

   I. $-3b < ba < 3b$

   II. $-a < a^2 < a \quad if \ a < 0$

   III. $-13 < 3a - 4 < 5$

A. I only

B. II only

C. I and III only

D. III only

E. I, II and III

7) Removing which of the following numbers will change the average of the numbers to 7?

$$2, 5, 6, 9, 12, 13$$

A. 12

B. 5

C. 6

D. 2

E. 13

8) A rope weighs 800 grams per meter of length. What is the weight in kilograms of 15.5 meters of this rope? (1 kilograms = 1,000 grams)

A. 0.0124

B. 0.124

C. 124

D. 1,240

E. 12,400

9) If $y = 2ab + 5b^2$, what is y when $a = 4$ and $b = 2$?

   A. 26
   B. 16
   C. 20
   D. 18
   E. 36

10) If $f(x) = 2 + 3x$ and $g(x) = -2x^2 - 6 - x$, then find $(g - f)(x)$?

   A. $2x^2 - 4x - 8$
   B. $2x^2 - 4x + 8$
   C. $-2x^2 - 4x + 8$
   D. $-2x^2 - 4x - 8$
   E. $-2x^2 + 4x - 8$

11) The marked price of a computer is D dollar. Its price decreased by 70% in January and later increased by 15 % in February. What is the final price of the computer in D dollar?

   A. 3.45 D
   B. 34.5 D
   C. 0.75 D
   D. 7.50 D
   E. 1.30 D

12) The number 64.7 is 100 times greater than which of the following numbers?

   A. 0.0647
   B. 0.647
   C. 0.00647
   D. 6.470
   E. 64.700

13) David's current age is 70 years, and Ava's current age is 10 years. In how many years David's age will be 5 times Ava's age?

   A. 5
   B. 7
   C. 9
   D. 12
   E. 15

14) How many tiles of 11 cm² is needed to cover a floor of dimension 7 cm by 33 cm?

　　A. 16　　　　C. 25　　　　E. 26

　　B. 19　　　　D. 21

15) What is the median of these numbers? 24, 9, 19, 32, 11, 15, 5

　　A. 15　　　　C. 19　　　　E. 32

　　B. 9　　　　D. 11

16) A company pays its employer $8,000 plus 6% of all sales profit. If x is the number of all sales profit, which of the following represents the employer's revenue?

　　A. $0.06x$　　　　C. $0.06x + 8,000$　　　　E. $-0.94x - 8,000$

　　B. $0.94x - 8,000$　　D. $0.94x + 8,000$

17) What is the area of a square whose diagonal is 6 cm?

　　A. 36 cm²　　　　C. 24 cm²　　　　E. 218 cm²

　　B. 18 cm²　　　　D. 12 cm²

18) What is the value of x in the following figure?

　　A. 125

　　B. 175

　　C. 55

　　D. 155

　　E. 105

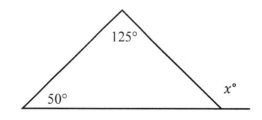

19) Right triangle ABC is shown below. Which of the following is true for all possible values of angle A and B?

A. $\cot A = \cot B$

B. $\tan^2 A = \tan^2 B$

C. $\cot A = 1$

D. $\cos A = \sin B$

E. $\cot A = \cos B$

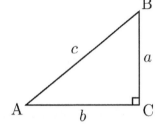

20) What is the value of y in the following system of equation?

$$2x - 5y = -8$$
$$-x + 2y = 3$$

A. -5

B. -3

C. 3

D. 6

E. 2

21) How long does a 312–miles trip take moving at 60 miles per hour (mph)?

A. 5 hours

B. 5 hours and 36 minutes

C. 5 hours and 18 minutes

D. 6 hours and 24 minutes

E. 12 hours and 45 minutes

22) From the figure, which of the following must be true? (figure not drawn to scale)

A. $y = 2z$

B. $y = 7x$

C. $y \geq x$

D. $y + 6x = z$

E. $z > x$

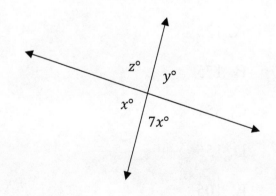

23) Which is the correct statement?

A. $\frac{1}{5} < 0.16$

B. $20\% = \frac{4}{5}$

C. $4 < \frac{7}{3}$

D. $\frac{5}{8} > 0.59$

E. None of them above

24) When 30% of 70 is added to 18% of 500, the resulting number is:

A. 21
B. 90
C. 111
D. 121
E. 189

25) A ladder leans against a wall forming a 60° angle between the ground and the ladder. If the bottom of the ladder is 30 feet away from the wall, how long is the ladder?

A. 40 feet
B. 80 feet
C. 50 feet
D. 60 feet
E. 160 feet

26) If 75% of a class are girls, and 12% of girls play tennis, what percent of the class play tennis?

A. 9%
B. 10%
C. 12%
D. 18%
E. 24%

27) If x is a real number, and if $x^3 + 52 = 170$, then x lies between which two consecutive integers?

A. 1 and 2
B. 2 and 3
C. 3 and 4
D. 4 and 5
E. 5 and 6

28) If $(x-5)^2 = 9$ which of the following could be the value of $(x-6)(x-5)$?

   A. 2   C. 7   E. −5

   B. 6   D. −2

29) Simplify.

$$4x^2 + 5y^5 - 2x^2 + 5z^3 - y^2 + 4x^3 - 2y^5 + 3z^3$$

A. $2x^2 - 4y^2 + 3y^5 + 8z^3$

B. $2x^2 + 4x^3 - y^2 + 3y^5 + 8z^3$

C. $2x^2 + 4x^3 + 3y^5 + 8z^3$

D. $2x^2 - 4x^3 - 2y^2 + y^5 + 8z^3$

E. $2x^2 + 4x^3 - 3y^2 + 8z^3$

30) In four successive hours, a car travels 44 km, 46 km, 42 km and 52 km. In the next four hours, it travels with an average speed of 55 km per hour. Find the total distance the car traveled in 8 hours.

   A. 404 km   C. 220 km   E. 1,440 km

   B. 184 km   D. 808 km

31) From last year, the price of gasoline has increased from $2.04 per gallon to $3.06 per gallon. The new price is what percent of the original price?

   A. 52%   C. 130%   E. 130%

   B. 150%   D. 170%

32) Simplify $(-2 + 3i)(5 + 6i)$,

   A. $28 - 3i$

   B. $8 - 3i$

   C. $-10 + 3i$

   D. $-28 + 3i$

   E. $3i$

33) If $\tan \theta = \frac{5}{12}$ and $\sin \theta > 0$, then $\cos \theta = ?$

   A. $-\frac{5}{12}$

   B. $\frac{12}{13}$

   C. $\frac{5}{13}$

   D. $-\frac{13}{5}$

   E. 1

34) Which of the following has the half period and five times the amplitude of graph $y = \sin x$?

   A. $y = \frac{1}{2} \sin 5x$

   B. $y = 5\sin(\frac{x}{2} + 5)$

   C. $y = 2.5 \sin 2x$

   D. $y = 5 + 5 \sin 2x$

   E. $y = 4 + \sin \frac{x}{2}$

35) Which of the following shows the numbers in increasing order?

   A. $\frac{1}{2}, \frac{8}{13}, \frac{5}{7}, \frac{2}{5}$

   B. $\frac{2}{5}, \frac{1}{2}, \frac{8}{13}, \frac{5}{7}$

   C. $\frac{5}{7}, \frac{2}{5}, \frac{8}{13}, \frac{1}{2}$

   D. $\frac{2}{5}, \frac{7}{11}, \frac{5}{7}, \frac{1}{2}$

   E. None of them above

36) In 1999, the average worker's income increased $4,000 per year starting from $36,000 annual salary. Which equation represents income greater than average? (I = income, x = number of years after 1999)

A. I > 4,000 x + 36,000

B. I > – 4,000 x + 36,000

C. I < –4,000 x + 36,000

D. I < 4,000 x – 36,000

E. I < 36,000 x + 4,000

Questions 37 to 39 are based on the following data

The result of a research shows the number of men and women in four cities of a country.

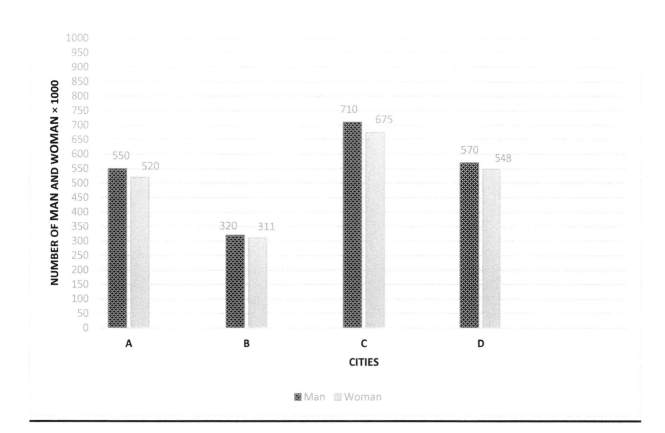

37) What's the ratio of percentage of men in city C to percentage of men in city B?

   A. 0.11   C. 0.99   E. 1.11

   B. 9.991   D. 99.11

38) What's the minimum ratio of woman to man in the four cities?

   A. 0.97   C. 0.95   E. 0.96

   B. 0.98   D. 0.99

39) How many women should be added to city D until the ratio of women to men will be 1.4?

A. 222   C. 128   E. 138

B. 228   D. 298

40) What are the values of mode and median in the following set of numbers?

5, 4, 8, 6, 8, 8, 5, 5, 5, 8, 4

A. Mode: 5, 4 Median: 4   D. Mode: 3, 5 Median: 4

B. Mode: 5, 8 Median: 5   E. Mode: 5, Median: 5

C. Mode: 2, 4 Median: 5

41) y is $x\%$ of what number?

A. $\frac{y}{100x}$ C. $\frac{100y}{x}$ E. $\frac{xy}{100}$

B. $\frac{x}{100y}$ D. $\frac{100x}{y}$

42) If cotangent of an angel β is $\sqrt{2}$, then the tangent of angle β is

A. -1 C. $\sqrt{2}$ E. $-\frac{\sqrt{2}}{2}$

B. 1 D. $\frac{\sqrt{2}}{2}$

43) If a box contains red and blue balls in ratio of 3: 8, how many red balls are there if 96 blue balls are in the box?

A. 45   C. 30   E. 11

B. 36   D. 20

44) 8 liters of water are poured into an aquarium that's 40cm long, 2cm wide, and 60cm high. How many cm will the water level in the aquarium rise due to this added water? (1 liter of water = 1,000 cm³)

A. 90   C. 40   E. 8

B. 100   D. 20

45) What is the surface area of the cylinder below?

A. 78 π in²

B. 87 π² in²

C. 98 π in²

D. 86 π² in²

E. 606 π in²

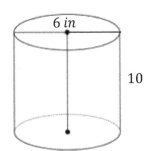

46) A chemical solution contains 4% alcohol. If there is 28ml of alcohol, what is the volume of the solution?

A. 350 ml   C. 700 ml   E. 1,200 ml

B. 450 ml   D. 1,100 ml

47) What is the solution of the following inequality?

$$|x - 5| \leq 2$$

A. $x \geq 7 \cup x \leq 3$ D. $x \leq 7$

B. $3 \leq x \leq 7$ E. Set of real numbers

C. $x \geq 10$

48) Which of the following points lies on the line $2x - 3y = 5$?

   A. $(1, -2)$

   B. $(-3, 0)$

   C. $(-2, -3)$

   D. $(1, -4)$

   E. $(0, -2)$

49) In the following figure, ABCD is a rectangle, and E and F are points on AD and DC, respectively. The area of $\triangle BED$ is 18, and the area of $\triangle BDF$ is 10. What is the perimeter of the rectangle?

   A. 32

   B. 24

   C. 18

   D. 56

   E. 48

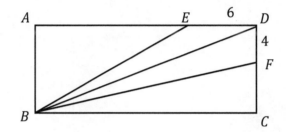

50) In the xy-plane, the point $(3, 9)$ and $(2, 8)$ are on line A. Which of the following equations of lines is parallel to line A?

   A. $y = 2x$

   B. $y = 8$

   C. $y = \frac{x}{3}$

   D. $y = -x$

   E. $y = x$

51) When point A $(7, 3)$ is reflected over the y-axis to get the point B, what are the coordinates of point B?

   A. $(7, 3)$

   B. $(-7, -3)$

   C. $(-7, 3)$

   D. $(7, -3)$

   E. $(0, 3)$

52) A bag contains 18 balls: three green, four black, six blue, a brown, two red and two white. If 17 balls are removed from the bag at random, what is the probability that a brown ball has been removed?

A. $\frac{1}{2}$

B. $\frac{1}{8}$

C. $\frac{18}{17}$

D. $\frac{1}{18}$

E. $\frac{17}{18}$

53) If a tree casts a 48–foot shadow at the same time that a 7 feet yardstick casts a 6–foot shadow, what is the height of the tree?

A. 13 ft

B. 52 ft

C. 56 ft

D. 48 ft

E. 42 ft

54) If the area of trapezoid is 210, what is the perimeter of the trapezoid?

A. 27

B. 62

C. 54

D. 58

E. 64

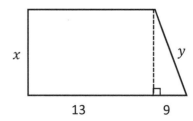

ACT Math Workbook

55) If 60% of x equal to 30% of 30, then what is the value of $(x + 3)^2$?

   A. 18.20   C. 32.04   E. 324

   B. 32   D. 3,240

56) If $f(x) = 2x^3 + 3x^2 + x$ and $g(x) = -2$, what is the value of $f(g(x))$?

   A. 6   C. 16   E. 7

   B. 2   D. -6

57) A boat sails 36 miles south and then 15 miles east. How far is the boat from its start point?

   A. 36 miles   C. 88 miles   E. 51 miles

   B. 39 miles   D. 38 miles

58) If $x \begin{bmatrix} 3 & 0 \\ 0 & 4 \end{bmatrix} = \begin{bmatrix} 2x + y - 5 & 0 \\ 0 & 3y - 12 \end{bmatrix}$, what is the product of x and y?

   A. 3   C. 8   E. 24

   B. 12   D. 20

59) If $f(x) = 5^x$ and $g(x) = \log_5 x$, which of the following expressions is equal to $f(5g(p))$?

   A. $5P$ C. p^5 E. $\frac{p}{5}$

   B. 5^p D. $5p^5$

60) In the following equation when z is divided by 6, what is the effect on x?

$$x = \frac{7y + \frac{r}{2r+3}}{\frac{12}{z}}$$

A. x is divided by 2

B. x is divided by 6

C. x does not change

D. x is multiplied by 6

E. x is multiplied by 2

STOP

This is the End of this Test. You may check your work on this Test if you still have time.

ACT Practice Test 2

Mathematics

- ❖ **60 Questions.**
- ❖ **Total time for this test: 60 Minutes**.
- ❖ **You may use a scientific calculator on this test.**

Administered *Month Year*

ACT Math Workbook

1) Convert 7,800,000 to scientific notation.

   A. $7.80 \times 1,000$

   B. 7.80×10^{-6}

   C. 7.8×100

   D. 7.8×10^6

   E. 7.8×10^5

2) $(x^3)^{\frac{4}{9}}$ equal to?

   A. $x^{\frac{4}{3}}$

   B. $x^{\frac{7}{9}}$

   C. $x^{\frac{4}{12}}$

   D. $x^{\frac{9}{13}}$

   E. $x^{\frac{4}{18}}$

3) Simplify $\frac{3-4i}{-3i}$?

   A. $\frac{4}{3} + i$

   B. $\frac{4}{3} - i$

   C. $\frac{1}{3} - i$

   D. $\frac{1}{3} + i$

   E. $3i$

4) What is the value of x in the following equation?

$$8^x = 512$$

   A. 8

   B. 3

   C. 8

   D. 9

   E. 2

5) What is the sum of prime numbers between 1 and 10?

   A. 4

   B. 10

   C. 15

   D. 17

   E. 18

ACT Math Workbook

6) If $\sqrt{7x} = \sqrt{y}$, then $x =$?

   A. $7y$

   B. $\sqrt{\frac{y}{7}}$

   C. y^2

   D. $\sqrt{7y}$

   E. $\frac{y}{7}$

7) The average weight of 24 girls in a class is 50 kg and the average weight of 26 boys in the same class is 55 kg. What is the average weight of all the 50 students in that class?

   A. 56

   B. 52.6

   C. 62.28

   D. 62.60

   E. 52

8) If $y = (-2x^4)^3$, which of the following expressions is equal to y?

   A. $-2x^4$

   B. $-2x^7$

   C. $8x^5$

   D. $8x^7$

   E. $-8x^{12}$

9) What is the value of the expression $4(x + y) + (2 - x)^2$ when $x = 3$ and $y = -1$?

   A. -12

   B. -21

   C. 9

   D. -9

   E. 16

10) Sophia purchased a sofa for $235.80 The sofa is regularly priced at $524. What was the percent discount Sophia received on the sofa?

   A. 1.65%

   B. 65%

   C. 55%

   D. 45%

   E. 1.45%

ACT Math Workbook

11) If $f(x) = 3x - 8$ and $g(x) = 2x^2 - 5x$, then find $\left(\frac{f}{g}\right)(x)$.

A. $\dfrac{3x-8}{2x^2-5x}$

B. $\dfrac{x-8}{2x^2-5x}$

C. $\dfrac{x-5}{x^2-4}$

D. $\dfrac{3x+8}{x^2+5x}$

E. $\dfrac{x^2-4x}{3x-8}$

12) In the standard (x, y) coordinate plane, which of the following lines contains the points $(1, -7)$ and $(6, 8)$?

A. $y = 3x - 10$

B. $y = \dfrac{1}{3}x + 10$

C. $y = -3x + 7$

D. $y = -\dfrac{1}{3}x + 10$

E. $y = 3x - 7$

13) A bank is offering 1.5% simple interest on a savings account. If you deposit $15,000, how much interest will you earn in three years?

A. $650

B. $675

C. $6,500

D. $6,700

E. $6,600

14) If the ratio of home fans to visiting fans in a crowd is 3: 5 and all 16,000 seats in a stadium are filled, how many visiting fans are in attendance?

A. 100,000

B. 1,000

C. 9,000

D. 10

E. 10,000

15) If the interior angles of a quadrilateral are in the ratio 1:3:7:9, what is the measure of the largest angle?

   A. 54°   C. 198°   E. 162°

   B. 18°   D. 126°

16) If $x + 2sin^2 a + 2cos^2 a = 6$, then $x =$?

   A. 2   C. 3   E. 7

   B. 4   D. 6

17) If the area of a circle is 100 square meters, what is its diameter?

   A. 10π C. $\frac{20\sqrt{\pi}}{\pi}$ E. $10\sqrt{\pi}$

   B. $\frac{10}{\pi}$

                       D. $100\pi^2$

18) The length of a rectangle is $\frac{4}{5}$ times its width. If the width is 40, what is the perimeter of this rectangle?

   A. 32   C. 144   E. 154

   B. 45   D. 121

19) In the figure below, line A is parallel to line B. What is the value of angle x?

   A. 55 degree

   B. 65 degree

   C. 95 degree

   D. 115 degree

   E. 125 degree

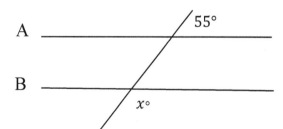

ACT Math Workbook

20) What is the value of *x* in the following system of equations?

$$x + 2y = 7$$
$$4x + 5y = 22$$

A. 3

B. 2

C. −3

D. 5

E. 6

21) An angle is equal to one fourth of its supplement. What is the measure of that angle?

A. 36

B. 26.5

C. 30

D. 45

E. 60

22) Last week 15,000 fans attended a football match. This week two times as many bought tickets, but one fifth of them cancelled their tickets. How many are attending this week?

A. 15,000

B. 6,000

C. 60,000

D. 24,000

E. 30,000

23) If $sin\alpha = \frac{\sqrt{3}}{2}$ in a right triangle and the angle α is an acute angle, then what is $cos\ \alpha$?

A. $\frac{\sqrt{3}}{3}$

B. $\frac{1}{3}$

C. $\sqrt{3}$

D. $\frac{1}{\sqrt{3}}$

E. $\frac{1}{2}$

24) In following **squares** which statement is true?

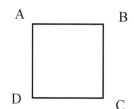

A. AB is parallel to BC

B. AB is perpendicular to DC

C. Length of AB equal to half of length BC

D. The measure of all the angles equals 360°.

E. The answer cannot be found from the information given.

25) In the standard (x, y) coordinate system plane, what is the area of the circle with the following equation?

$$(x + 2)^2 + (y - 1)^2 = 9$$

A. 2π

B. 9π

C. 18π

D. 6π

E. 3π

26) In two successive years, the population of a town is increased by 15% and 30%. What percent of the population is increased after two years?

A. 30%

B. 45%

C. 49.5%

D. 35.5%

E. 55.5%

27) Simplify.

$$8x^5y^2 + 3x^3y^4 - (2x^5y^2 - 4x^3y^4)$$

A. $-x^5y^3$

B. $6x^5y^3 - 7x^5y^5$

C. $8x^2y^3$

D. $6x^5y^2 + 7x^3y^4$

E. $8x^5y^6$

28) What are the zeroes of the function $f(x)= 2x^3+ 14x^2+ 20x$?

A. 0

B. $-2, 4$

C. $0, 2, 4$

D. $-2, -3$

E. $0, -2, -5$

29) If one angle of a right triangle measures 30°, what is the sine of the other acute angle?

A. $\frac{\sqrt{3}}{2}$

B. $\frac{\sqrt{2}}{2}$

C. $\frac{1}{2}$

D. 1

E. $\sqrt{2}$

30) In the following figure, what is the perimeter of $\triangle ABC$ if the area of $\triangle ADC$ is 45?

A. 57.5

B. 26

C. 25

D. 60

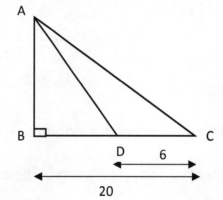

E. The answer cannot be determined from the information given

31) Which of the following is one solution of this equation?

$$3x^2 + 5x - 8 = 0$$

A. $\sqrt{5} + 1$

B. $\sqrt{5} - 1$

C. 1

D. $\sqrt{5}$

E. $\sqrt{15}$

ACT Math Workbook

32) Three-kilograms apple and four-kilograms orange cost $51.2. If one-kilogram apple costs $2.4 how much does one-kilogram orange cost?

A. $11

C. $6.5

E. $12

B. $8

D. $7

33) Which of the following expressions is equal to $\sqrt{\frac{3x^2}{5} + \frac{x^2}{25}}$?

A. $5x$

C. $2x\sqrt{x}$

E. $4x$

B. $\frac{4x}{5}$

D. $\frac{x\sqrt{x}}{5}$

Questions 34 to 36 are based on the following data

34) Between which two of the months shown was there a twenty percent increase in the number of pants sold?

   A. January and February

   B. February and March

   C. March and April

   D. April and May

   E. May and June

35) During the six-month period shown, what is the mean number of shirts and median number of shoes per month?

   A. 30, 137.5

   B. 149, 25

   C. 139, 25

   D. 30, 139

   E. 30, 25

36) How many shoes need to be added in February until the ratio of number of pants to number of shoes in February equals to seven-eighth of this ratio in March?

   A. 2
   B. 9
   C. 5
   D. 18
   E. 20

37) A card is drawn at random from a standard 52–card deck, what is the probability that the card is of hearts or diamonds? (The deck includes 13 of each suit clubs, diamonds, hearts, and spades)

   A. $\frac{1}{5}$
   B. $\frac{1}{2}$
   C. $\frac{1}{13}$
   D. $\frac{1}{26}$
   E. $\frac{1}{52}$

38) A football team had $25,000 to spend on supplies. The team spent $18,000 on new balls. New sport shoes cost $110 each. Which of the following inequalities represent how many new shoes the team can purchase?

   A. $110x + 18{,}000 \leq 25{,}000$
   B. $110x + 18{,}000 \geq 25{,}000$
   C. $25{,}000x + 110 \leq 18{,}000$
   D. $18{,}000x + 110 \geq 25{,}000$
   E. $18{,}000x + 25{,}000 \geq 110$

39) If $x = 4$, what is the value of y in the following equation? $5y = \frac{3x^2}{8} + 9$

   A. 3
   B. 15
   C. 65
   D. 12
   E. 120

40) A swimming pool holds 4,800 cubic feet of water. The swimming pool is 15 feet long and 8 feet wide. How deep is the swimming pool?

A. 8 feet
C. 15 feet
E. 40 feet

B. 23 feet
D. 25 feet

41) The ratio of boys to girls in a school is 3:7. If there are 320 students in a school, how many boys are in the school.

A. 80
C. 320
E. 96

B. 600
D. 116

42) If $(x-2)^2 + 3 > 2x + 3$, then x can equal which of the following?

A. 1
C. 7
E. 3

B. 5
D. 2

43) Let r and p be constants. If $x^2 + 4x + r$ factors into $(x+3)(x+p)$, the values of r and p respectively are?

A. 3, 1
C. 2, 3

B. 1, 3
D. 3, 2

E. The answer cannot be found from the information given.

44) If 140% of a number is 70, then what is 80% of that number?

A. 35
C. 40
E. 70

B. 60
D. 45

45) The width of a box is half of its length. The height of the box is half of its width. If the length of the box is 20 cm, what is the volume of the box?

   A. 200 cm³

   B. 100 cm³

   C. 2,000 cm³

   D. 1,000 cm³

   E. 10,000 cm³

46) The average of six consecutive numbers is 24. What is the smallest number?

   A. 25

   B. 30.5

   C. 21.5

   D. 15.5

   E. 15

47) The surface area of a cylinder is $120\pi\ cm^2$. If its height is 7 cm, what is the radius of the cylinder?

   A. 18 cm

   B. 10 cm

   C. 12 cm

   D. 5 cm

   E. 4 cm

48) In a coordinate plane, triangle ABC has coordinates: $(5, -1)$, $(-4, -2)$, and $(2, 4)$. If triangle ABC is reflected over the y-axis, what are the coordinates of the new image?

   A. $(5, -1), (-2, -4), (4, 2)$

   B. $(-2, 4), (-4, -2), (5, -1)$

   C. $(-5, -1), (4, -2), (-2, 4)$

   D. $(4, -2), (-2, 4), (5, -2)$

   E. $(-4, -2), (2, 4), (-5, -1)$

49) What is the slope of a line that is perpendicular to the line $8x - 2y = 16$?

   A. -4

   B. $-\frac{1}{4}$

   C. 6

   D. 8

   E. 16

50) What is the difference in area between a 8 cm by 4 cm rectangle and a circle with diameter of 12 cm? ($\pi = 3$)

   A. 46

   B. 76

   C. 32

   D. 108

   E. 12

51) If $f(x)=3x^3+5$ and $g(x)=\frac{2}{x}$, what is the value of $f(g(x))$?

   A. $\frac{8}{5x^3+5}$

   B. $\frac{5}{x^3}$

   C. $\frac{6}{5x}$

   D. $\frac{1}{5x+5}$

   E. $\frac{32}{x^3}+5$

52) A cruise line ship left Port A and traveled 21 miles due west and then 28 miles due north. At this point, what is the shortest distance from the cruise to port A?

   A. 50 miles

   B. 40 miles

   C. 30 miles

   D. 35 miles

   E. 49 miles

53) The length of a rectangle is 5 meters greater than 7 times its width. The perimeter of the rectangle is 90 meters. What is the area of the rectangle?

   A. 45 m²

   B. 300 m²

   C. 200 m²

   D. 90 m²

   E. 180 m²

54) Tickets to a movie cost $10.50 for adults and $5.50 for students. A group of 18 friends purchased tickets for $119. How many student tickets did they buy?

A. 4
B. 14
C. 11
D. 18
E. 7

55) What is the solution of the following inequality?

$$|x - 4| \geq 7$$

A. $x \geq 11 \cup x \leq -3$
B. $-3 \leq x \leq 11$
C. $x \geq 11$
D. $x \leq -3$
E. Set of real numbers

56) If $\tan x = \frac{15}{20}$, then $\sin x =$

A. $\frac{1}{5}$
B. $\frac{15}{25}$
C. $\frac{12}{25}$
D. $\frac{7}{25}$
E. It cannot be determined from the information given.

57) In the following figure, ABCD is a rectangle. If $a = \sqrt{2}$, and $b = 2a$, find the area of the shaded region. (the shaded region is a trapezoid)

A. 15
B. 10
C. $12\sqrt{2}$
D. $8\sqrt{2}$
E. $4\sqrt{2}$

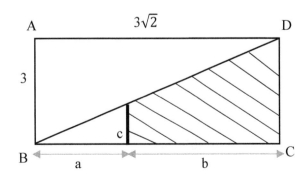

ACT Math Workbook

58) If the ratio of $9a$ to $8b$ is $\frac{1}{16}$, what is the ratio of a to b?

   A. 16

   B. 18

   C. $\frac{1}{18}$

   D. $\frac{1}{32}$

   E. $\frac{1}{16}$

59) If $A = \begin{bmatrix} 1 & 1 \\ 2 & -1 \end{bmatrix}$ and $B = \begin{bmatrix} 4 & 2 \\ -2 & 3 \end{bmatrix}$, then $3A - B =$

   A. $\begin{bmatrix} -3 & 1 \\ 4 & -4 \end{bmatrix}$

   B. $\begin{bmatrix} -7 & 1 \\ 2 & -6 \end{bmatrix}$

   C. $\begin{bmatrix} 4 & 2 \\ -1 & 3 \end{bmatrix}$

   D. $\begin{bmatrix} 1 & 5 \\ 0 & -1 \end{bmatrix}$

   E. $\begin{bmatrix} -7 & 1 \\ 8 & -6 \end{bmatrix}$

60) What is the amplitude of the graph of the equation $y - 2 = 5\cos 2x$? (half the distance between the graph's minimum and maximum y-values in standard (x, y) coordinate plane is the amplitude of a graph.)

   A. 2

   B. 5

   C. 4

   D. 2.5

   E. 0.3

STOP

This is the End of this Test. You may check your work on this Test if you still have time.

Answers and Explanations

ACT Math Practice Tests

Answer Key

❋ Now, it's time to review your results to see where you went wrong and what areas you need to improve!

ACT Math Practice Test

Practice Test 1						Practice Test 2					
1	C	21	C	41	C	1	D	21	A	41	E
2	D	22	D	42	D	2	A	22	D	42	C
3	B	23	D	43	B	3	A	23	E	43	A
4	E	24	C	44	B	4	B	24	D	44	C
5	C	25	D	45	A	5	D	25	B	45	D
6	C	26	A	46	C	6	E	26	C	46	C
7	A	27	D	47	B	7	B	27	D	47	D
8	C	28	B	48	C	8	E	28	E	48	C
9	E	29	B	49	A	9	C	29	A	49	B
10	D	30	A	50	E	10	B	30	D	50	B
11	B	31	B	51	C	11	A	31	C	51	E
12	B	32	D	52	D	12	A	32	A	52	D
13	A	33	B	53	C	13	B	33	B	53	C
14	D	34	D	54	B	14	E	34	D	54	B
15	A	35	B	55	E	15	E	35	C	55	A
16	C	36	A	56	D	16	B	36	B	56	B
17	B	37	D	57	B	17	C	37	B	57	E
18	B	38	C	58	E	18	C	38	A	58	C
19	D	39	B	59	C	19	E	39	A	59	E
20	E	40	B	60	B	20	A	40	E	60	B

Answers and Explanations
ACT Mathematics
Practice Tests 1

1) Answer: C.

$5^{\frac{5}{2}} \times 5^{\frac{1}{2}} = 5^{\frac{5}{2}+\frac{1}{2}} = 5^{\frac{6}{2}} = 5^3$

2) Answer: D.

Solve for x, $\frac{5x}{36} = \frac{x-2}{6}$

Multiply the second fraction by 6, $\frac{5x}{36} = \frac{6(x-2)}{6 \times 6}$

Tow denominators are equal. Therefore, the numerators must be equal.

$5x = 6x - 12 \rightarrow -x = -12 \rightarrow x = 12$

3) Answer: B.

Simplify each option provided.

A. $25 - (5 \times 8) + (7 \times 20) = 25 - 40 + 140 = 125$

B. $\left(\frac{10}{7} \times 63\right) + \left(\frac{124}{4}\right) = 90 + 31 = 121$ (this is the answer)

C. $\left(\left(\frac{70}{6} + \frac{22}{3}\right) \times 5\right) - \frac{20}{3} + \frac{130}{6} = \left(\left(\frac{70+44}{6}\right) \times 5\right) - \frac{40}{6} + \frac{130}{6} = \left(\left(\frac{114}{6}\right) \times 5\right) + \frac{130-40}{6} = (19 \times 5) + \frac{90}{6} = 95 + 15 = 110$

D. $(3 \times 11) + (42 \times 2.5) - 14 = 33 + 105 - 14 = 124$

E. $\frac{148}{8} + \frac{207}{2} = \frac{148+828}{8} = 122$

4) Answer: E.

five years ago, Amy was three times as old as Mike. Mike is 11 years now. Therefore, 5 years ago Mike was 6 years.

five years ago, Amy was: $A = 3 \times 6 = 18$

Now Amy is 23 years old: $18 + 5 = 23$

ACT Math Workbook

5) Answer: C.

Set of number that are not composite between 1 and 15: A = {2, 3, 5, 7, 11, 13}

Probability $= \frac{number\ of\ desired\ outcomes}{number\ of\ total\ outcomes} = \frac{6}{15} = \frac{2}{5}$

6) Answer: C.

I. $|a| < 3 \to -3 < a < 3$

Multiply all sides by b. Since, $b > 0 \to -3b < ba < 3b$ (it is true!)

II. Since, $-3 < a < 3$, and $a < 0 \to -a > a^2 > a$ (plug in $-\frac{1}{3}$, and check!) (It's false)

III. $-3 < a < 3$, multiply all sides by 3, then: $-9 < 3a < 9$

Subtract 4 from all sides. Then:

$-9 - 4 < 3a - 4 < 9 - 4 \to -13 < 3a - 4 < 5$ (It is true!)

7) Answer: A.

Check each option provided:

A. 12 $\frac{2+5+6+9+13}{5} = \frac{35}{5} = 7$

B. 5 $\frac{2+6+9+12+13}{5} = \frac{42}{5} = 8.4$

C. 6 $\frac{2+5+9+12+13}{5} = \frac{41}{5} = 8.2$

D. 2 $\frac{5+6+12+9+13}{5} = \frac{45}{5} = 9$

E. 13 $\frac{2+5+6+9+12}{5} = \frac{34}{5} = 6.8$

8) Answer: C.

The weight of 15.5 meters of this rope is: 15.5 × 800 g = 124,000 g

1 kg = 1,000 g, therefore, 124,000 g ÷ 1,000 = 124 kg

9) Answer: E.

$y = 2ab + 5b^2$

Plug in the values of a and b in the equation: $a = 4$ and $b = 2$

$y = 2(4)(2) + 5(2)^2 = 16 + 5(4) = 16 + 20 = 36$

WWW.MathNotion.Com

ACT Math Workbook

10) Answer: D.

$(g-f)(x) = g(x) - f(x) = (-2x^2 - 6 - x) - (2 + 3x)$

$-2x^2 - 6 - x - 2 - 3x = -2x^2 - 4x - 8$

11) Answer: B.

To find the discount, multiply the number by (100% – rate of discount).

Therefore, for the first discount we get: (D) (100% – 70%) = (D) (0.30) = 0.30 D

For increase of 15 %: (0.30 D) (100% + 15%) = (0.30 D) (1.15) = 0.345 D = 34.5% of D

12) Answer: B.

100 times the number is 64.7. Let x be the number, then:

$100x = 64.7 \rightarrow x = \frac{64.7}{100} = 0.647$

13) Answer: A.

Let's review the options provided.

A. 5. In 5 years, David will be 75 and Ava will be 15. 75 is 5 times 15.

B. 7. In 7 years, David will be 77 and Ava will be 17. 77 is NOT 5 times 17.

C. 9. In 9 years, David will be 79 and Ava will be 19. 79 is not 5 times 19.

D. 12. In 12 years, David will be 82 and Ava will be 22. 82 is not 5 times 22.

E. 15. In 15 years, David will be 85 and Ava will be 25. 85 is not 5 times 25.

14) Answer: D.

The area of the floor is: 7 cm × 33cm = 231 cm

The number is tiles needed = 231 ÷ 11 = 21

15) Answer: A.

Write the numbers in order:

5, 9, 11, 15, 19, 24, 32

Since we have 7 numbers (7 is odd), then the median is the number in the middle, which is 15.

16) Answer: C.

Employer's revenue: $0.6x + 8,000$

17) Answer: B.

The diagonal of the square is 6. Let x be the side.

Use Pythagorean Theorem: $a^2 + b^2 = c^2$

$x^2 + x^2 = 6^2 \Rightarrow 2x^2 = 36 \Rightarrow 2x^2 = 36 \Rightarrow x^2 = 18 \Rightarrow x = \sqrt{18}$

The area of the square is: $\sqrt{18} \times \sqrt{18} = 18$

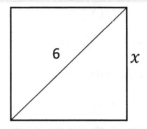

18) Answer: B.

$x = 50 + 125 = 175$

19) Answer: D.

By definition, the sine of any acute angle is equal to the cosine of its complement.

Since, angle A and B are complementary angles, therefore:

$\cos A = \sin B$

20) Answer: E.

Solve the system of equations by elimination method.

$\begin{array}{l} 2x - 5y = -8 \\ -x + 2y = 3 \end{array}$ Multiply the second equation by 2, then add it to the first equation.

$\begin{array}{l} 2x - 5y = -8 \\ 2(-x + 2y = 3) \end{array} \Rightarrow \begin{array}{l} 2x - 5y = -8 \\ -2x + 4y = 6) \end{array} \Rightarrow$ add the equations, $y = 2$

21) Answer: C.

Use distance formula:

Distance = Rate × time $\Rightarrow 312 = 60 \times T$, divide both sides by 40.

$\frac{312}{60} = T \Rightarrow T = 5.2$ hours.

Change hours to minutes for the decimal part. 0.2 hours = $0.2 \times 60 = 18$ minutes.

22) Answer: D.

x and z are colinear. y and $6x$ are colinear. Therefore,

$x + z = y + 7x$, subtract x from both sides, then, $z = y + 6x$

23) Answer: D.

Check each option.

A. $\frac{1}{5} < 0.16$ $\frac{1}{4} = 0.20$ and it is less than 0.16. Not true!

B. $20\% = \frac{2}{10}$ $20\% = \frac{1}{5} < \frac{4}{5}$. Not True!

C. $4 < \frac{7}{3}$ $\frac{7}{3} = 2.333 < 4$. Not True!

D. $\frac{5}{8} > 0.625$ $\frac{5}{8} = 0.625$ and it is greater than 0.59. Bingo!

E. None of them above   Not True!

24) Answer: C.

30% of 70 equals to: $0.30 \times 70 = 21$

18% of 500 equals to: $0.18 \times 500 = 90$

30% of 70 is added to 18% of 500: $21 + 90 = 111$

25) Answer: D.

The relationship among all sides of special right triangle

$30°, 60°, 90°$ is provided in this triangle:

In this triangle, the opposite side of $30°$ angle is half of the

hypotenuse. Draw the shape of this question.

The ladder is the hypotenuse.

Therefore, the ladder is 60 ft.

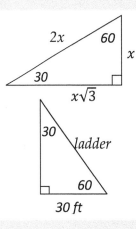

26) Answer: A.

The percent of girls playing tennis is: $75\% \times 12\% = 0.75 \times 0.12 = 0.09 = 9\%$

27) Answer: D.

Solve for x. $x^3 + 52 = 170 \Rightarrow x^3 = 118$

Let's review the options.

A. 1 and 2. $1^3 = 1$ and $2^3 = 8$, 118 is not between these two numbers.

B. 2 and 3. $2^3 = 8$ and $3^3 = 27$, 118 is not between these two numbers.

C. 3 and 4. $3^3 = 27$ and $4^3 = 64$, 118 is not between these two numbers.

D. 4 and 5. $4^3 = 64$ and $5^3 = 125$, 118 is between these two numbers.

E. 5 and 6. $5^3 = 125$ and $6^3 = 216$, 118 is not between these two numbers.

ACT Math Workbook

28) Answer: B.

$(x-5)^2 = 9 \to x - 5 = 3 \to x = 8$

$\to (x-6)(x-5) = (8-6)(8-5) = (2)(3) = 6$

29) Answer: B.

$4x^2 + 5y^5 - 2x^2 + 5z^3 - y^2 + 4x^3 - 2y^5 + 3z^3 = 4x^2 - 2x^2 + 4x^3 - y^2 + 5y^5 - 2y^5 + 5z^3 + 3z^3 = 2x^2 + 4x^3 - y^2 + 3y^5 + 8z^3$

30) Answer: A.

Add the first 4 numbers. $44 + 46 + 42 + 52 = 184$

To find the distance traveled in the next 4 hours, multiply the average by number of hours.

Distance = Average × Rate = $55 \times 4 = 220$

Add both numbers. $220 + 184 = 404$

31) Answer: B.

The question is this: 3.06 is what percent of 2.04?

Use percent formula: part = $\frac{\text{percent}}{100}$ × whole

$3.06 = \frac{\text{percent}}{100} \times 2.04 \Rightarrow 3.06 = \frac{\text{percent} \times 2.04}{100} \Rightarrow 306 = \text{percent} \times 2.04 \Rightarrow \text{percent} = \frac{306}{2.04} = 150$

32) Answer: D.

We know that: $i = \sqrt{-1} \Rightarrow i^2 = -1$

$(-2 + 3i)(5 + 6i) = -10 - 12i + 15i + 18i^2 = -10 + 3i - 18 = 3i - 28$

33) Answer: B.

$tan\theta = \frac{\text{opposite}}{\text{adjacent}}$

$tan\theta = \frac{5}{12} \Rightarrow$ we have the following right triangle. Then,

$c = \sqrt{5^2 + 12^2} = \sqrt{25 + 144} = \sqrt{169} = 13$

$cos\theta = \frac{\text{adjacent}}{\text{hypotenuse}} = \frac{12}{13}$

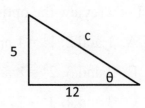

34) Answer: D.

The amplitude in the graph of the equation $y = a sin bx$ is a. (a and b are constant)

ACT Math Workbook

In the equation $y = sinx$, the amplitude is 1 and the period of the graph is 2π.

The only option that has five times the amplitude of graph $y = \sin x$ is $y = 5 + 5\sin 2x$ for the half period $sin2x = sin2\pi \Rightarrow 2x = 2\pi \Rightarrow x = \pi$

They both have the amplitude of 5 and period of π.

35) Answer: B.

$\frac{2}{5} = 0.4$, $\frac{1}{2} = 0.5$, $\frac{8}{13} \cong 0.61$, $\frac{5}{7} \cong 0.71$

36) Answer: A.

Let x be the number of years. Therefore, $4,000 per year equals $4,000x$.

starting from $36,000 annual salary means you should add that amount to $4,000x$.

Income more than that is: $I > 4,000x + 36,000$.

37) Answer: D.

Percentage of men in city C = $\frac{710}{1,385} \times 100 = 51.26\%$

Percentage of men in city B = $\frac{320}{631} \times 100 = 50.71\%$

Percentage of men in city C to percentage of men in city B: $\frac{51.26}{50.71} = 99.11$

38) Answer: C.

Ratio of women to men in city A: $\frac{520}{550} = 0.95$

Ratio of women to men in city B: $\frac{311}{320} = 0.97$

Ratio of women to men in city C: $\frac{675}{710} = 0.95$

Ratio of women to men in city D: $\frac{548}{570} = 0.96$

39) Answer: B.

Let the number of women should be added to city D be x, then:

$\frac{548 + x}{570} = 1.4 \rightarrow 548 + x = 570 \times 1.4 = 798 \rightarrow x = 228$

40) Answer: B.

We write the numbers in the order: 4, 4, 5, 5, 5, 5, 6, 8, 8, 8, 8

The mode of numbers is: 5 and 8, median is: 5

ACT Math Workbook

41) Answer: C.

Let the number be A. Then: $y = x\% \times A \rightarrow$ (Solve for A) $\rightarrow x = \frac{x}{100} \times A$

Multiply both sides by $\frac{100}{x}$: $y \times \frac{100}{x} = \frac{x}{100} \times \frac{100}{x} \times A \rightarrow A = \frac{100y}{x}$

42) Answer: D.

$tangent\ \beta = \frac{1}{cotangent\ \beta} = \frac{1}{\sqrt{2}} = \frac{\sqrt{2}}{2}$

43) Answer: B.

$\frac{3}{8} \times 96 = 36$

44) Answer: B.

One liter = $1,000 cm^3 \rightarrow 8$ liters = $8,000\ cm^3$

$8,000 = 40 \times 2 \times h \rightarrow h = \frac{8,000}{80} = 100$ cm

45) Answer: A.

Surface Area of a cylinder = $2\pi r\ (r + h)$,

The radius of the cylinder is 3 (6÷2) inches and its height is 10 inches. Therefore,

Surface Area of a cylinder = $2\pi\ (3)\ (3 + 10) = 78\ \pi$

46) Answer: C.

4% of the volume of the solution is alcohol. Let x be the volume of the solution.

Then: 4% of $x = 28ml \Rightarrow 0.04\ x = 28 \Rightarrow x = 28 \div 0.04 = 700$

47) Answer: B.

$|x - 5| \leq 2 \rightarrow -2 \leq x - 5 \leq 2 \rightarrow -2 + 5 \leq x - 5 + 5 \leq 2 + 5 \rightarrow 3 \leq x \leq 7$

48) Answer: C.

Plug in each pair of number in the equation:

A. $(1, -2)$: $2(1) - 3(-2) = 8$ Nope!

B. $(-3, 0)$: $2(-3) - 3(0) = -6$ Nope!

C. $(-2, -3)$: $2(-2) - 3(-3) = 5$ Bingo!

D. $(1, -4)$: $2(1) - 3(-4) = 14$ Nope!

E. $(0, -2)$: $2(0) - 3(-2) = 6$ Nope!

WWW.MathNotion.Com

49) Answer: A.

The area of ΔBED is 18, then: $\frac{6 \times AB}{2} = 18 \rightarrow 6 \times AB = 36 \rightarrow AB = 6$

The area of ΔBDF is 10, then: $\frac{2 \times BC}{2} = 10 \rightarrow 2 \times BC = 20 \rightarrow BC = 10$

The perimeter of the rectangle is $= 2 \times (6 + 10) = 32$

50) Answer: E.

The slop of line A is: $m = \frac{y_2 - y_1}{x_2 - x_1} = \frac{9-8}{3-2} = 1$

Parallel lines have the same slope and only choice E ($y = x$) has slope of 1.

51) Answer: C.

When points are reflected over y-axis, the value of y in the coordinates doesn't change and the sign of x changes. Therefore, the coordinates of point B is $(-7, 3)$.

52) Answer: D.

If 18 balls are removed from the bag at random, there will be one ball in the bag. The probability of choosing a brown ball is 1 out of 17. Therefore, the probability of not choosing a brown ball is 17 out of 18 and the probability of having not a brown ball after removing 17 balls is the same.

53) Answer: C.

Write a proportion and solve for x.

$\frac{7}{6} = \frac{x}{48} \Rightarrow 6x = 7 \times 48 \Rightarrow x = 56$ ft

54) Answer: B.

The area of trapezoid is: $\left(\frac{13+22}{2}\right) \times x = 210 \rightarrow 17.5x = 210 \rightarrow x = 12$

$y = \sqrt{9^2 + 12^2} = 15$

Perimeter is: $22 + 12 + 13 + 15 = 62$

55) Answer: E.

$0.6x = (0.3) \times 30 \rightarrow x = 15 \rightarrow (x+3)^2 = (18)^2 = 324$

56) Answer: D.

$g(x) = -2,$

ACT Math Workbook

then $f(g(x)) = f(-2) = 2(-2)^3 + 3(-2)^2 + (-2) = -16 + 12 - 2 = -6$

57) Answer: B.

Use the information provided in the question to draw the shape.

Use Pythagorean Theorem: $a^2 + b^2 = c^2$

$15^2 + 36^2 = c^2 \Rightarrow 225 + 1{,}296 = c^2$

$\Rightarrow 1{,}521 = c^2 \Rightarrow c = 39$

58) Answer: E.

$\begin{cases} 3x = 2x + y - 5 \\ 4x = 3y - 12 \end{cases} \rightarrow \begin{cases} x - y = -5 \\ 4x - 3y = -12 \end{cases}$

Multiply first equation by -4.

$\begin{cases} -4x + 4y = 20 \\ 4x - 3y = -12 \end{cases} \rightarrow$ add two equations.

$y = 8 \rightarrow y = 8 \rightarrow x = 3 \rightarrow x \times y = 24$

59) Answer: C.

To solve for $f(5g(p))$, first, find $5g(p)$

$g(x) = \log_5 x \rightarrow g(p) = \log_5 p \rightarrow 5g(p) = 5\log_5 p = \log_5 p^5$

Now, find $f(5g(p))$: $f(x) = 5^x \rightarrow f(\log_5 p^5) = 5^{\log_5 p^5}$

Logarithms and exponentials with the same base cancel each other. This is true because logarithms and exponentials are inverse operations. Then: $f(\log_5 p^5) = 5^{\log_5 p^5} = p^5$

60) Answer: B.

$x_1 = \dfrac{7y + \frac{r}{2r+3}}{\frac{z}{6}} = \dfrac{7y + \frac{r}{2r+3}}{\frac{6 \times 12}{z}} = \dfrac{7y + \frac{r}{2r+3}}{6 \times \frac{12}{z}} = \dfrac{1}{6} \times \dfrac{7y + \frac{r}{2r+3}}{\frac{12}{z}} = \dfrac{x}{6}$

Answers and Explanations
ACT Mathematics
Practice Tests 2

1) **Answer: D.**

   $7,800,000 = 7.8 \times 10^6$

2) **Answer: A.**

   $(x^3)^{\frac{4}{9}} = x^{3 \times \frac{4}{9}} = x^{\frac{12}{9}} = x^{\frac{4}{3}}$

3) **Answer: A.**

   To simplify the fraction, multiply both numerator and denominator by i.

   $\frac{3-4i}{-3i} \times \frac{i}{i} = \frac{3i-4i^2}{-3i^2}$

   $i^2 = -1$, Then: $\frac{3i-4i^2}{-3i^2} = \frac{3i-4(-1)}{-3(-1)} = \frac{3i+4}{3} = \frac{3i}{3} + \frac{4}{3} = i + \frac{4}{3}$

4) **Answer: B.**

   $512 = 8^3 \rightarrow 8^x = 8^3 \rightarrow x = 3$

5) **Answer: D.**

   Here is the list of all prime numbers between 1 and 10:

   2, 3, 5, 7

   The sum of all prime numbers between 1 and 10 is:

   $2 + 3 + 5 + 7 = 17$

6) **Answer: E.**

   Solve for x. $\sqrt{7x} = \sqrt{y}$

   Square both sides of the equation: $(\sqrt{7x})^2 = (\sqrt{y})^2$

   $7x = y \rightarrow x = \frac{y}{7}$

7) **Answer: B.**

   $\text{average} = \frac{\text{sum of terms}}{\text{number of terms}}$

   The sum of the weight of all girls is: $24 \times 50 = 1,200$ kg

ACT Math Workbook

The sum of the weight of all boys is: $26 \times 55 = 1{,}430$ kg

The sum of the weight of all students is: $1{,}200 + 1{,}430 = 2{,}630$ kg

average $= \frac{2{,}630}{50} = 52.6$

8) Answer: E.

$y = (-2x^4)^3 = (-2)^3 (x^4)^3 = -8x^{12}$

9) Answer: C.

Plug in the value of x and y. $x = 3$ and $y = -1$

$4(x + y) + (2 - x)^2 = 4(3 + (-1)) + (2 - 3)^2 = 4(3 - 1) + (-1)^2 = 9$

10) Answer: B.

The question is this: 235.80 is what percent of 524?

Use percent formula: part $= \frac{\text{percent}}{100} \times$ whole

$235.80 = \frac{\text{percent}}{100} \times 524 \Rightarrow 235.80 = \frac{\text{percent} \times 524}{100} \Rightarrow 23{,}580 = \text{percent} \times 524 \Rightarrow \text{percent}$

$= \frac{23{,}580}{524} = 45$

235.80 is 45 % of 524. Therefore, the discount is: 100% − 45% = 65%

11) Answer: A.

$\left(\frac{f}{g}\right)(x) = \frac{f(x)}{g(x)} = \frac{3x - 8}{2x^2 - 5x}$

12) Answer: A.

The equation of a line is: $y = mx + b$, where m is the slope and b is the y-intercept.

First find the slope: $m = \frac{y_2 - y_1}{x_2 - x_1} = \frac{8 - (-7)}{6 - 1} = \frac{15}{5} = 3$

Then, we have: $y = 3x + b$

Choose one point and plug in the values of x and y in the equation to solve for b.

Let's choose the point $(1, -7)$

$y = 3x + b \rightarrow -7 = 3(1) + b \rightarrow -7 = 3 + b \rightarrow b = -10$

The equation of the line is: $y = 3x - 10$

ACT Math Workbook

13) Answer: B.

Use simple interest formula:

$I = prt$ (I = interest, p = principal, r = rate, t = time)

$I = (15,000)(0.015)(3) = 675$

14) Answer: E.

Number of visiting fans: $\frac{5 \times 16,000}{8} = 10,000$

15) Answer: E.

The sum of all angles in a quadrilateral is 360 degrees.

Let x be the smallest angle in the quadrilateral. Then the angles are: $x, 3x, 7x, 9x$

$x + 3x + 7x + 9x = 360 \rightarrow 20x = 360 \rightarrow x = 18$

The angles in the quadrilateral are: 18°, 54°, 126°, and 162°

16) Answer: B.

$2\sin^2 a + 2\cos^2 a = 2(\sin^2 a + \cos^2 a) = 2(1) = 2$, then:

$x + 2 = 6 \rightarrow x = 4$

17) Answer: C.

Formula for the area of a circle is: $A = \pi r^2$

Using 100 for the area of the circle we have: $100 = \pi r^2$

Let's solve for the radius (r).

$\frac{100}{\pi} = r^2 \rightarrow r = \sqrt{\frac{100}{\pi}} = \frac{10}{\sqrt{\pi}} = \frac{10}{\sqrt{\pi}} \times \frac{\sqrt{\pi}}{\sqrt{\pi}} = \frac{10\sqrt{\pi}}{\pi} \rightarrow d = 2r = 2 \times \frac{10\sqrt{\pi}}{\pi} \rightarrow d = \frac{20\sqrt{\pi}}{\pi}$

18) Answer: C.

Length of the rectangle is: $\frac{4}{5} \times 40 = 32$

perimeter of rectangle is: $2 \times (32 + 40) = 144$

19) Answer: E.

The angle x and 45 are complementary angles. Therefore:

$x + 55 = 180$

$180° - 55° = 125°$

20) Answer: A.

Solving Systems of Equations by Elimination

Multiply the first equation by (–4), then add it to the second equation.

$\begin{array}{l}-4(x+2y=7)\\ 4x+5y=22\end{array} \Rightarrow \begin{array}{l}-4x-8y=-28\\ 4x+5y=22\end{array} \Rightarrow -3y=-6 \Rightarrow y=2$

Plug in the value of y into one of the equations and solve for x.

$x + 2(2) = 7 \Rightarrow x + 4 = 7 \Rightarrow x = 7 - 4 \Rightarrow x = 3$

21) Answer: A.

The sum of supplement angles is 180. Let x be that angle. Therefore, $x + 4x = 180 \Rightarrow$
$5x = 180$, divide both sides by 5: $x = 36$

22) Answer: D.

Two times of 15,000 is 30,000. One fifth of them cancelled their tickets.

One fifth of 30,000 equal 6,000 ($\frac{1}{5} \times 30,000 = 6,000$).

24,000 (30,000 – 6,000 = 24,000) fans are attending this week

23) Answer: E.

$sin\alpha = \frac{\sqrt{3}}{2} \Rightarrow$ Since $sin\alpha = \frac{opposite}{hypotenuse}$, we have the following right triangle. Then,

$c = \sqrt{2^2 - (\sqrt{3})^2} = \sqrt{4-3} = \sqrt{1} = 1$

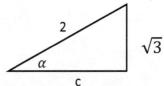

$cos\alpha = \frac{1}{2}$

24) Answer: D.

In any squares measure of all angles equals 360°.

25) Answer: B.

The equation of a circle in standard form is:

$(x - h)^2 + (y - k)^2 = r^2$, where r is the radius of the circle.

In this circle the radius is 3. $r^2 = 9 \rightarrow r = 3$

$(x + 2)^2 + (y - 1)^2 = 3^2$

Area of a circle: $A = \pi r^2 = \pi(3)^2 = 9\pi$

26) Answer: C.

the population is increased by 15% and 30%. 15% increase changes the population to 115% of original population.

For the second increase, multiply the result by 130%.

$(1.15) \times (1.30) = 1.495 = 149.5\%$

49.5 percent of the population is increased after two years.

27) Answer: D.

$8x^5y^2 + 3x^3y^4 - (2x^5y^2 - 4x^3y^4) = 8x^5y^2 - 2x^5y^2 + 3x^3y^4 + 4x^3y^4 = 6x^5y^2 + 7x^3y^4$

28) Answer: E.

Frist factor the function: $f(x) = 2x^3 + 14x^2 + 20x = 2x(x+2)(x+5)$

To find the zeros, $f(x)$ should be zero. $f(x) = 2x(x+2)(x+5) = 0$

Therefore, the zeros are: $x = 0$

$(x+2) = 0 \Rightarrow x = -2$; $(x+5) = 0 \Rightarrow x = -5$

29) Answer: A.

The relationship among all sides of right triangle 30°, 60°, 90° is provided in the following triangle:

Sine of 60° equals to: $\frac{opposite}{hypotenuse} = \frac{x\sqrt{3}}{2x} = \frac{\sqrt{3}}{2}$

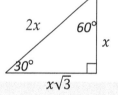

30) Answer: D.

Let x be the length of AB, then: $45 = \frac{x \times 6}{2} \rightarrow x = 15$

The length of $AC = \sqrt{15^2 + 20^2} = \sqrt{625} = 25$

The perimeter of $\triangle ABC = 15 + 20 + 25 = 60$

31) Answer: C.

$x_{1,2} = \frac{-b \pm \sqrt{b^2 - 4ac}}{2a}$

$ax^2 + bx + c = 0 \Rightarrow 3x^2 + 5x - 8 = 0$, then: a = 3, b = 5 and c = − 8

$x = \frac{-5 + \sqrt{5^2 - 4 \times 3 \times (-8)}}{2 \times 3} = 1$; $x = \frac{-5 - \sqrt{5^2 - 4 \times 3 \times (-8)}}{2 \times 3} = -\frac{8}{3}$

ACT Math Workbook

32) Answer: A.

Let x be the cost of one-kilogram orange, then:

$4x + (3 \times 2.4) = 51.2 \rightarrow 4x + 7.2 = 51.2 \rightarrow 4x = 51.2 - 7.2 \rightarrow 4x = 44 \rightarrow x = \frac{44}{4} = \$11$

33) Answer: B.

Simplify the expression.

$\sqrt{\frac{3x^2}{5} + \frac{x^2}{25}} = \sqrt{\frac{15x^2}{25} + \frac{x^2}{25}} = \sqrt{\frac{16x^2}{25}} = \sqrt{\frac{16}{25}x^2} = \sqrt{\frac{16}{25}} \times \sqrt{x^2} = \frac{4}{5} \times x = \frac{4x}{5}$

34) Answer: D.

First find the number of pants sold in each month.

January: 100, February: 78, March: 80, April: 60, May: 74, June: 55

Check each option provided.

A. There is a decrease from January to February

B. February and March,

$\left(\frac{80-78}{80}\right) \times 100 = \frac{2}{80} \times 100 = 2.5\%$

C. There is a decrease from March to April

D. April and May: there is an increase from April to May

$\left(\frac{74-60}{74}\right) \times 100 = \frac{14}{74} \times 100 = 18.92\%$

E. There is a decrease from May to June.

35) Answer: C.

First, order the number of shirts sold each month:

$120, 130, 135, 140, 150, 160$

mean is: $\frac{120+130+135+140+150+160}{6} = \frac{834}{6} = 139$

Put the number of shoes sold per month in order:

$15, 20, 20, 30, 30, 35$; median is: $\frac{20+30}{2} = 25$

36) Answer: B.

The ratio of number of pants to number of shoes in March equals $\frac{80}{35}$.

Seven-eighth of this ratio is $\left(\frac{7}{8}\right)\left(\frac{80}{35}\right)$. Now, let x be the number of shoes needed to be added in February.

$\frac{78}{30+x} = \left(\frac{7}{8}\right)\left(\frac{80}{35}\right) \to \frac{78}{30+x} = \frac{560}{280} = 2 \to 78 = 2(30+x) \to 78 = 60 + 2x \to 2x = 18 \to x = 9$

37) **Answer: B.**

The probability of choosing a heart or diamonds is $\frac{26}{52} = \frac{1}{2}$

38) **Answer: A.**

Let x be the number of shoes the team can purchase. Therefore, the team can purchase $110\,x$.

The team had $25,000 and spent $18,000. Now the team can spend on new shoes $7,000 at most.

Now, write the inequality: $110x + 18,000 \le 25,000$

39) **Answer: A.**

Plug in the value of x in the equation and solve for y.

$5y = \frac{3x^2}{8} + 9 \to 5y = \frac{3(4)^2}{8} + 9 \to 5y = \frac{3(16)}{8} + 9 \to 5y = 6 + 9 = 15$

$\to 5y = 15 \to y = 3$

40) **Answer: E.**

Use formula of rectangle prism volume.

V = (length) (width) (height) $\Rightarrow$ 4,800 = (8) (15) (height) $\Rightarrow$ height = 4,800 ÷ 120 = 40

41) **Answer: E.**

The ratio of boy to girls is 3:7. Therefore, there are 3 boys out of 10 students. To find the answer, first divide the total number of students by 10, then multiply the result by 3.

320 ÷ 10 = 32 $\Rightarrow$ 32 × 3 = 96

42) **Answer: C.**

Plug in the value of each option in the inequality.

A. 1 $(1-2)^2 + 3 > 2(1) + 3 \to 4 > 5$ No!

B. 5 $(5-2)^2 + 3 > 2(5) + 3 \to 12 > 13$ No!

C. 7	$(7-2)^2 + 3 > 2(7) + 3 \to 28 > 17$		Bingo!
D. 2	$(2-2)^2 + 3 > 2(2) + 3 \to 3 > 7$		No!
E. 3	$(3-2)^2 + 3 > 2(3) + 3 \to 4 > 9$		No!

43) Answer: A.

$(x+3)(x+p) = x^2 + (3+p)x + 3p \to 3 + p = 4 \to p = 1$ and $r = 3p = 3$

44) Answer: C.

First, find the number.

Let x be the number. Write the equation and solve for x.

140% of a number is 70, then:

$1.4 \times x = 70 \Rightarrow x = 70 \div 1.4 = 50$

80% of 50 is: $0.8 \times 50 = 40$

45) Answer: D.

If the length of the box is 20, then the width of the box is half of it, 10, and the height of the box is 5 (half of the width). The volume of the box is:

V = (length) × (width) × (height) = (20) × (10) × (5) = 1,000

46) Answer: C.

Let x be the smallest number. Then, these are the numbers:

$x, x+1, x+2, x+3, x+4, x+5$

average $= \frac{\text{sum of terms}}{\text{number of terms}} \Rightarrow 24 = \frac{x+(x+1)+(x+2)+(x+3)+(x+4)+(x+5)}{5} \Rightarrow 24 = \frac{6x+15}{6} \Rightarrow 144$

$= 6x + 15 \Rightarrow 129 = 6x \Rightarrow x = 21.5$

47) Answer: D.

Formula for the Surface area of a cylinder is:

$SA = 2\pi r^2 + 2\pi rh \to 120\pi = 2\pi r^2 + 2\pi r(7) \to r^2 + 7r - 60 = 0$

Factorize and solve for r.

$(r+12)(r-5) = 0 \to r = 5$ or $r = -12$ (unacceptable)

48) Answer: C.

Since the triangle *ABC* is reflected over the *y*-axis, then all values of *y*'s of the points don't change and the sign of all *x*'s change.

(remember that when a point is reflected over the *y*-axis, the value of *y* does not change and when a point is reflected over the *x*-axis, the value of *x* does not change).

Therefore:

$(5, -1)$ changes to $(-5, -1)$

$(-4, -2)$ changes to $(4, -2)$

$(2, 4)$ changes to $(-2, 4)$

49) Answer: B.

The equation of a line in slope intercept form is: $y = mx + b$

Solve for y.

$8x - 2y = 16 \Rightarrow -2y = 16 - 8x \Rightarrow y = (16 - 8x) \div (-2) \Rightarrow$

$y = 4x - 8 \rightarrow$ The slope is 4.

The slope of the line perpendicular to this line is:

$m_1 \times m_2 = -1 \Rightarrow 4 \times m_2 = -1 \Rightarrow m_2 = -\frac{1}{4}$

50) Answer: B.

The area of rectangle is: $8 \times 4 = 32 \ cm^2$

The area of circle is: $\pi r^2 = \pi \times (\frac{12}{2})^2 = 3 \times 36 = 108 \ cm^2$

Difference of areas is: $108 - 32 = 76$

51) Answer: E.

$f(g(x)) = 4 \times (\frac{2}{x})^3 + 5 = \frac{32}{x^3} + 5$

52) Answer: D.

Use the information provided in the question to draw the shape. Use Pythagorean Theorem: $a^2 + b^2 = c^2$

$21^2 + 28^2 = c^2 \Rightarrow 441 + 784 = c^2$

$\Rightarrow 1,225 = c^2 \Rightarrow c = 35$

ACT Math Workbook

53) Answer: C.

Let L be the length of the rectangular and W be the with of the rectangular. Then, $L = 7W + 5$

The perimeter of the rectangle is 90 meters. Therefore:
$$2L + 2W = 90$$
$$L + W = 45$$

Replace the value of L from the first equation into the second equation and solve for W:
$$(7W + 5) + W = 45 \to 8W + 5 = 45 \to 8W = 40 \to W = 5$$

The width of the rectangle is 4 meters and its length is:
$$L = 7W + 5 = 7(5) + 5 = 40$$

The area of the rectangle is: length × width = 40 × 5 = 200

54) Answer: B.

Let x be the number of adult tickets and y be the number of student tickets. Then:

$x + y = 18$

$10.50x + 5.50y = 119$

Use elimination method to solve this system of equation. Multiply the first equation by -5.5 and add it to the second equation.

$-5.5(x + y = 18) \Rightarrow -5.5x - 5.5y = -99$

$10.50x + 5.50y = 119 \Rightarrow 5x = 20 \to x = 4$

There are 4 adults' tickets and 14 student tickets.

55) Answer: A.

$x - 4 \geq 7 \to x \geq 7 + 4 \to x \geq 11$

Or $x - 4 \leq -7 \to x \leq -7 + 4 \to x \leq -3$

Then, solution is: $\quad x \geq 11 \ \cup \ x \leq -3$

56) Answer: B.

$\tan = \frac{opposite}{adjacent}$, and $\tan x = \frac{15}{20}$, therefore, the opposite side of the angle x is 15 and the adjacent side is 20. Let's draw the triangle.

Using Pythagorean theorem, we have:

$a^2 + b^2 = c^2 \rightarrow 15^2 + 20^2 = c^2 \rightarrow 225 + 400 = c^2 \rightarrow c = 25$

$\sin x = \frac{opposite}{hypotenuse} = \frac{15}{25} = \frac{3}{5}$

57) Answer: E.

Based on triangle similarity theorem:

$\frac{a}{a+b} = \frac{c}{3} \rightarrow c = \frac{3a}{a+b} = \frac{3\sqrt{2}}{\sqrt{2}+2\sqrt{2}} = 1$

$\rightarrow$ area of shaded region is: $\left(\frac{c+3}{2}\right)(b) = \frac{4}{2} \times 2\sqrt{2} = 4\sqrt{2}$

58) Answer: C.

Write the ratio of $9a$ to $8b$, $\frac{9a}{8b} = \frac{1}{16}$

Use cross multiplication and then simplify.

$9a \times 16 = 8b \times 1 \rightarrow 144a = 8b \rightarrow a = \frac{8b}{144} = \frac{b}{18}$

Now, find the ratio of a to b.

$\frac{a}{b} = \frac{\frac{b}{18}}{b} \rightarrow \frac{b}{18} \div b = \frac{b}{18} \times \frac{1}{b} = \frac{b}{18b} = \frac{1}{18}$

59) Answer: E.

First, find $3A$.

$A = \begin{bmatrix} 1 & 1 \\ 2 & -1 \end{bmatrix} \Rightarrow 3A = 3 \times \begin{bmatrix} 1 & 1 \\ 2 & -1 \end{bmatrix} = \begin{bmatrix} 3 & 3 \\ 6 & -3 \end{bmatrix}$

Now, solve for $3A - B$:

$\begin{bmatrix} -3 & 3 \\ 6 & -3 \end{bmatrix} - \begin{bmatrix} 4 & 2 \\ -2 & 3 \end{bmatrix} = \begin{bmatrix} -3-4 & 3-2 \\ 6-(-2) & -3-3 \end{bmatrix} = \begin{bmatrix} -7 & 1 \\ 8 & -6 \end{bmatrix}$

60) Answer: B.

The amplitude in the graph of the equation $y = a\cos bx$ is a. (a and b are constant)

In the equation $y - 2 = 5\cos 2x$, the amplitude is 5.

"End"

CPSIA information can be obtained
at www.ICGtesting.com
Printed in the USA
LVHW061120140223
739386LV00011B/137